# BOB MILLER'S
# MATH PREP FOR THE
# COMPASS® EXAM

**Bob Miller**
Former Lecturer in Mathematics
City College of New York
New York, NY

 *Res*

*Research & Education Association*
61 Ethel Road West
Piscataway, New Jersey 08854
E-mail: info@rea.com

**BOB MILLER'S**
# Math Prep for the COMPASS® Exam

Library of Congress Control Number 2011934272

ISBN-13: 978-0-7386-1002-3
ISBN-10: 0-7386-1002-X

# CONTENTS

# AUTHOR ACKNOWLEDGMENTS

I have many people to thank.

I thank my wife, Marlene, who makes life worth living, who is truly the wind under my wings.

I thank the rest of my family: children Sheryl and Eric and their spouses Glenn and Wanda (who are also like my children); grandchildren Kira, Evan, Sean, Sarah, Ethan, and Noah; my brother Jerry; and my parents, Cele and Lee; and my in-law parents, Edith and Siebeth.

I thank Larry Kling, Alice Leonard, and Mel Friedman for making this book possible.

I thank Martin Levine for making my whole writing career possible.

I have been negligent in thanking my great math teachers of the past. I thank Mr. Douglas Heagle, Mr. Alexander Lasaka, Mr. Joseph Joerg, and Ms. Arloeen Griswold, the best math teacher I ever had, of George W. Hewlett High School; Ms. Helen Bowker of Woodmere Junior High; and Professor Pinchus Mendelssohn and Professor George Bachman of New York City Polytechnic. The death of Professor Bachman was an extraordinary loss to our country, which produces too few advanced degrees in math. Every year, two or three of Professor Bachman's students would receive a Ph.D. in math, and even more would receive their M.S. in math. In addition, he wrote four books and numerous papers on subjects that had never been written about or had been written so poorly that nobody could understand the material. His teachings and writings are clear and memorable.

As usual, the last three thanks go to three terrific people: a great friend, Gary Pitkofsky; another terrific friend and fellow lecturer, David Schwinger; and my cousin, Keith Robin Ellis, the sharer of our dreams.

*Bob Miller*

## DEDICATION

*To my wife, Marlene. I dedicate this book and everything else I ever do to you.*
*I love you very, very much.*

# AUTHOR BIOGRAPHY

I received my B.S. in the Unified Honors Program sponsored by the Ford Foundation and my M.S. in math from Polytechnic Institute of NYU. After teaching my first class, as a substitute for a full professor, I heard one student say to another upon leaving the classroom, "At least we have someone who can teach the stuff." I was hooked forever on teaching. Since then, I have taught at C.U.N.Y., Westfield State College, Rutgers, and Poly. No matter how I feel, I always feel a lot better when I teach. I always feel great when students tell me they used to hate math or couldn't do math and now they like it more and can do it better.

My main blessing is my family. I have a fantastic wife in Marlene. My kids are wonderful: daughter Sheryl, son Eric, son-in-law Glenn, and daughter-in-law Wanda. My grandchildren are terrific: Kira, Evan, Sean, Sarah, Ethan, and Noah. My hobbies are golf, bowling, bridge, crossword puzzles, and Sudoku. My ultimate goals are to write a book to help parents teach their kids math, a high school text that will advance our kids' math abilities, and a calculus text students can actually understand.

To me, teaching is always a great joy. I hope that I can give some of that joy to you. I know that this book will help you get the score you need to get into the proper math class at the college of your choice.

# OTHER BOOKS

Bob Miller's Math for the Accuplacer

Bob Miller's Math for the ACT

Bob Miller's Math for the GMAT, Second Edition

Bob Miller's Math for the GRE, Third Edition

Bob Miller's Math for the TABE

Bob Miller's Basic Math and Pre-Algebra for the Clueless, Second Edition

Bob Miller's Algebra for the Clueless, Second Edition

Bob Miller's Geometry for the Clueless, Second Edition

Bob Miller's Math SAT for the Clueless, Second Edition

Bob Miller's Pre-Calc with Trig for the Clueless, Third Edition

Bob Miller's High School Calc for the Clueless

Bob Miller's Calc 1 for the Clueless, Second Edition

Bob Miller's Calc 2 for the Clueless, Second Edition

Bob Miller's Calc 3 for the Clueless

## ABOUT RESEARCH & EDUCATION ASSOCIATION

Founded in 1959, Research & Education Association (REA) is dedicated to publishing the finest and most effective educational materials—including software, study guides, and test preps—for students in elementary school, middle school, high school, college, graduate school, and beyond.

Today, REA's wide-ranging catalog is a leading resource for teachers, students, and professionals.

We invite you to visit us at *www.rea.com* to find out how "REA is making the world smarter."

# TO THE STUDENT: A MUST READ

Congratulations!!!! You are about to begin college, a new and exciting time in your life. You are preparing to take the COMPASS®, an untimed test that is designed to properly place you in college. In math you will be asked questions in five areas: numerical skills and pre-algebra, college algebra (pre-calculus), geometry, and trigonometry.

The COMPASS® has on its website seven suggestions. Here is my own list.

1. Relax! It is important to do to come to the exam well-rested. I suggest you do not study the day before the test. Last-minute cramming rarely helps and usually hurts.

2. Get a good night's sleep before the test. Eat a good breakfast. Get to the test early so you are in a good frame of mind. If you are a slow starter, practice a few easy math problems before the test—any problems. Dress comfortably. Make sure you bring in everything you need including an acceptable fully-charged calculator. Bring a light snack in to keep up your energy.

3. Be sure you understand the directions. Ask if you do not.

4. Since this test is untimed, read a question over until you understand it.

5. You should answer every item since you are not penalized for wrong guessing. Since the test is computer adaptive, you *must* answer every question.

6. Don't be afraid to change an answer. *However,* I believe you should change an answer only if you are 100% (not 99%) sure the new answer is correct. I will use me as an example. The last math test I ever took I changed two answers. The first one I must have misread. I was 100% sure the new answer was correct and it was. The other I wasn't sure, and changed a correct answer to a wrong one!!!!

7. If you have a problem or question during the text, ask the administrator or proctor to help, remembering they cannot answer the question for you.

All of the questions on the test are multiple-choice. In the book some of the questions are not. Why? Sometimes you can figure out an answer from the choices. You want to practice so that you really know the material.

You may find this book is more than a test prep. This book actually reviews most of the topics needed for pre-calculus and calculus. If you have any weaknesses, this book will really help to correct them. You may find some topics written in ways to clear up problems you had or didn't know you had in high school math.

Good luck in college and in your future!

*Bob Miller*

## REA ACKNOWLEDGMENTS

In addition to our author, we would like to thank Larry B. Kling, Vice President, Editorial, for his overall direction; Pam Weston, Publisher, for setting the quality standards for production integrity and managing the publication to completion; Alice Leonard, Senior Editor, for project management; Mel Friedman, Lead Mathematics Editor, for editorial contributions; Kathleen Casey, Senior Editor, for preflight editorial review; and Transcend Creative Services (TCS), for typesetting this edition. Back-cover photo of author by Eric L. Miller.

"*You have a 60-inch ribbon. If you cut it so that each piece is 6 inches, how many cuts do you make?*"

**Our** great adventure begins with our basic terms. It is very important to understand what a question asks as well as how to answer it. The word *numbers* has many meanings, as we start to see here.

## NUMBERS

**Whole numbers:** 0, 1, 2, 3, 4, . . .

**Integers:** 0, $\pm 1$, $\pm 2$, $\pm 3$, $\pm 4$, . . . , where $\pm 3$ stands for both $+3$ and $-3$.

**Positive integers** are integers that are greater than 0. In symbols, $x > 0$, $x$ an integer.

**Negative integers** are integers that are less than 0. In symbols, $x < 0$, $x$ an integer.

**Even integers:** 0, $\pm 2$, $\pm 4$, $\pm 6$, . . .

**Odd integers:** $\pm 1$, $\pm 3$, $\pm 5$, $\pm 7$, . . .

### Inequalities

For any numbers represented by *a*, *b*, *c*, or *d* on the number line:

We say $c > d$ (*c* is greater than *d*) if *c* is to the right of *d* on the number line.

We say $d < c$ (*d* is less than *c*) if *d* is to the left of *c* on the number line.

$c > d$ is equivalent to $d < c$.

$a \leq b$ means $a < b$ or $a = b$; likewise, $a \geq b$ means $a > b$ or $a = b$.

**Example 1:**   $4 \leq 7$ is true because $4 < 7$; $9 \leq 9$ is true because $9 = 9$; but $7 \leq 2$ is false because $7 > 2$.

**Example 2:**   Find all integers between $-4$ and 5.

**Solution:**    $\{-3, -2, -1, 0, 1, 2, 3, 4\}$.

Notice that the word *between* does *not* include the endpoints.

**Example 3:**   Graph all the multiples of five between 20 and 40 inclusive.

**Solution:**    

Notice that *inclusive* means to include the endpoints.

## Odd and Even Numbers

Here are some facts about odd and even integers that you should know.

- The sum of two even integers is even.
- The sum of two odd integers is even.
- The sum of an even integer and an odd integer is odd.
- The product of two even integers is even.
- The product of two odd integers is odd.
- The product of an even integer and an odd integer is even.
- If $n$ is even, $n^2$ is even. If $n^2$ is even and $n$ is an integer, then $n$ is even.
- If $n$ is odd, $n^2$ is odd. If $n^2$ is odd and $n$ is an integer, then $n$ is odd.

## OPERATIONS ON NUMBERS

**Product** is the answer in multiplication, **quotient** is the answer in division, **sum** is the answer in addition, and **difference** is the answer in subtraction.

Because $3 \times 4 = 12$, 3 and 4 are said to be **factors** or **divisors** of 12, and 12 is both a **multiple** of 3 and a **multiple** of 4.

A **prime** is a positive integer with exactly two distinct factors, itself and 1. The number 1 is not a prime because only $1 \times 1 = 1$. It might be a good idea to memorize the first eight primes:

2, 3, 5, 7, 11, 13, 17, and 19

The number 4 has more than two factors: 1, 2, and 4. Numbers with more than two factors are called **composites**. The number 28 is a **perfect** number because if we add the factors less than 28, they add to 28.

**Example 4:** Write all the factors of 28.

**Solution:** 1, 2, 4, 7, 14, and 28.

**Example 5:** Write 28 as the product of prime factors.

**Solution:** $28 = 2 \times 2 \times 7$.

**Example 6:** Find all the primes between 70 and 80.

**Solution:** 71, 73, 79. How do we find this easily? First, because 2 is the only even prime, we have to check only the odd numbers. Next, we have to know the divisibility rules:

- A number is divisible by 2 if it ends in an even number. We don't need this here because then it can't be prime.

- A number is divisible by 3 (or 9) if the sum of the digits is divisible by 3 (or 9). For example, 456 is divisible by 3 because the sum of the digits is 15, which is divisible by 3 (it's not divisible by 9, but that's okay).

- A number is divisible by 4 if the number named by the last two digits is divisible by 4. For example, 3936 is divisible by 4 because 36 is divisible by 4.

- A number is divisible by 5 if the last digit is 0 or 5.

- The rule for 6 is a combination of the rules for 2 and 3.

- It is easier to divide by 7 than to learn the rule for 7.

- A number is divisible by 8 if the number named by the last *three* digits is divisible by 8.

- A number is divisible by 10 if it ends in a zero, as you know.

- A number is divisible by 11 if the difference between the sum of the even-place digits (2nd, 4th, 6th, etc.) and the sum of the odd-place digits (1st, 3rd, 5th, etc.) is a multiple of 11. For example, for the number 928,193,926: the sum of the odd-place digits (9, 8, 9, 9, and 6) is 41; the sum of the even-place digits (2, 1, 3, and 2) is 8; and $41 - 8$ is 33, which is divisible by 11. So 928,193,926 is divisible by 11.

That was a long digression!!!!! Let's get back to Example 6.

We have to check only 71, 73, 75, 77, and 79. The number 75 is not a prime because it ends in a 5. The number 77 is not a prime because it is divisible by 7. To see if the other three are prime, for any number less than 100 you have to divide by the primes 2, 3, 5, and 7 only. You will quickly find that 71, 73, and 79 are primes.

## Rules for Operations on Numbers

 *( ) are called parentheses (singular: parenthesis); [ ] are called brackets; { } are called braces.*

### Rules for adding signed numbers

1.   If all the signs are the same, add the numbers and use that sign.

2.   If two signs are different, subtract them, and use the sign of the larger numeral.

> **Example 7:**   **a.** $3 + 7 + 2 + 4 = +16$       **c.** $5 - 9 + 11 - 14 = 16 - 23 = -7$
>
> **b.** $-3 - 5 - 7 - 9 = -24$       **d.** $2 - 6 + 11 - 1 = 13 - 7 = +6$

### Rules for multiplying and dividing signed numbers
Look at the minus signs only.

1.   Odd number of minus signs—the answer is minus.

2.   Even number of minus signs—the answer is plus.

> **Example 8:**   $\dfrac{(-4)(-2)(-6)}{(-2)(+3)(-1)} =$
>
> **Solution:**    Five minus signs, so the answer is minus, namely $-8$.

### Rule for subtracting signed numbers
The sign $(-)$ means subtract. Change the problem to an addition problem.

> **Example 9:**   **a.** $(-6) - (-4) = (-6) + (+4) = -2$
>
> **b.** $(-6) - (+2) = (-6) + (-2) = -8$, because it is now an adding problem.

## Order of Operations

In doing a problem such as $4 + 5 \times 6$, the **order of operations** tells us whether to multiply or add first:

1.   If given letters, substitute in parentheses the value of each letter.

2.   Do operations in parentheses, inside ones first, and then the tops and bottoms of fractions.

3.   Do exponents next. (Chapter 3 discusses exponents in more detail.)

4. Do multiplications and divisions, left to right, as they occur.

5. The last step is adding and subtracting. Left to right is usually the safest way.

**Example 10:** $4 + 5 \times 6 =$

**Solution:**   $4 + 30 = 34$

**Example 11:** $(4 + 5)6 =$

**Solution:**   $(9)(6) = 54$

**Example 12:** $1000 \div 2 \times 4 =$

**Solution:**   $(500)(4) = 2000$

**Example 13:** $1000 \div (2 \times 4) =$

**Solution:**   $1000 \div 8 = 125$

**Example 14:** $4[3 + 2(5 - 1)] =$

**Solution:**   $4[3 + 2(4)] = 4[3 + 8] = 4(11) = 44$

**Example 15:** $\dfrac{3^4 - 1^{10}}{4 - 10 \times 2} =$

**Solution:**   $\dfrac{81 - 1}{4 - 20} = \dfrac{80}{-16} = -5$

**Example 16:** If $x = -3$ and $y = -4$, find the value of:

   a. $7 - 5x - x^2$

   b. $xy^2 - (xy)^2$

   c. $\dfrac{y^2 - x^3 - 3}{4 - 2xy}$

**Solutions:**   a. $7 - 5x - x^2 = 7 - 5(-3) - (-3)^2 = 7 + 15 - 9 = 13$

   b. $xy^2 - (xy)^2 = (-3)(-4)^2 - ((-3)(-4))^2 = (-3)(16) - (12)^2$

   $= -48 - 144 = -192$

   c. $\dfrac{y^2 - x^3 - 3}{4 - 2xy} = \dfrac{(-4)^2 - (-3)^3 - 3}{4 - 2(-3)(-4)} = \dfrac{16 + 27 - 3}{4 - 24} = \dfrac{40}{-20} = -2$

Before we get to the exercises, let's talk about ways to describe a group of numbers (data).

# DESCRIBING DATA

Four of the measures that describe data are necessary to know. The first three are measures of central tendency; the fourth, the range, measures the span of the data.

**Mean:**  Add up the numbers and divide by how many numbers you have added up.

**Median:**  Middle number. Put the numbers in numeric order and see which one is in the middle. If there are two "middle" numbers, which happens with an even number of data points, take their average.

**Mode:**  Most common numbers—those that appear the most times. A set with two modes is called bimodal. There can actually be any number of modes.

**Range:**  Highest number minus the lowest number.

**Example 17:**  Find the mean, median, mode, and range for 5, 6, 9, 11, 12, 12, and 14.

**Solutions:**  Mean:  $\dfrac{5 + 6 + 9 + 11 + 12 + 12 + 14}{7} = \dfrac{69}{7} = 9\dfrac{6}{7}$

Median:  11

Mode:  12

Range:  $14 - 5 = 9$

**Example 18:**  Find the mean, median, mode, and range for 4, 4, 7, 10, 20, 20.

**Solutions:**  Mean:  $\dfrac{4 + 4 + 7 + 10 + 20 + 20}{6} = \dfrac{65}{6} = 10\dfrac{5}{6}$

Median:  For an even number of points, it is the mean of the middle two:

$\dfrac{7 + 10}{2} = 8.5$

Mode:  There are two: 4 and 20 (blackbirds?)

Range: $20 - 4 = 16$

**Example 19:**    Jim received grades of 83 and 92 on two tests. What grade must the third test be in order to have an average (mean) of 90?

**Solution:**    There are two solution methods.

Method 1: To get a 90 average on three tests, Jim needs $3(90) = 270$ points. So far, he has $83 + 92 = 175$ points. So Jim needs $270 - 175 = 95$ points on the third test.

Method 2 (my favorite): 83 is $-7$ from 90. 92 is $+2$ from 90. $-7 + 2 = -5$ from the desired 90 average. Jim needs $90 + 5 = 95$ points on the third test. (Jim needs to "make up" the 5-point deficit, so add it to the average of 90.)

The second method is my choice, but it is always your choice which method you can do easier and faster.

**Q**    **Finally, after a long introduction, we get to some multiple-choice exercises.**

**Exercise 1:**    If $x = -5$, the value of $-3 - 4x - x^2$ is

A. $-48$         D. 13

B. $-8$          E. 4

C. 2

**Exercise 2:**    $-0(2) - \dfrac{0}{2} - 2 =$

A. 0             D. $-6$

B. $-2$          E. Undefined

C. $-4$

**Exercise 3:**    The scores on three tests were 90, 91, and 98. What does the score on the fourth test have to be in order to get exactly a 95 average (mean)?

A. 97            D. 100

B. 98            E. Not possible

C. 99

**Exercise 4:**   On a true-false test, 20 students scored 90, and 30 students scored 100. The sum of the mean, median, and mode is

A. 300              D. 294

B. 296              E. 275

C. 295

**Exercise 5:**   On a test, $m$ students received a grade of $x$, $n$ students received a grade of $y$, and $p$ students received a grade of $z$. The average (mean) grade is:

A. $\dfrac{mxnypz}{xyz}$          D. $\dfrac{mx + ny + pz}{m + n + p}$

B. $\dfrac{mx + ny + pz}{x + y + z}$          E. $mnp + \dfrac{xyz}{x + y + z}$

C. $\dfrac{mx + ny + pz}{xyz}$

**Exercise 6:**   The largest positive integer in the following list that divides evenly into 2,000,000,000,000,003 is

A. 33              D. 3

B. 11              E. 1

C. 10

For Exercises 7–9, use the following numbers: 8, 10, 10, 16, 16, 18

**Exercise 7:**   The mean is

A. 8              D. 16

B. 10              E.  There are two of them.

C. 13

**Exercise 8:**   The median is

A. 8              D. 16

B. 10              E.  There are two of them.

C. 13

**Exercise 9:**    The mode is

A. 8                     D. 16

B. 10                    E.  There are two of them.

C. 13

Sometimes statistics are given in frequency distribution tables, such as this one showing the grades Sandy received on 10 English quizzes.

### Sandy's Quiz Scores

| Grade | Number |
|-------|--------|
| 100 | 4 |
| 98 | 3 |
| 95 | 2 |
| 86 | 1 |
| Total | 10 |

This chart is for Exercises 10–12.

**Exercise 10:**    The mean is

A. 96                    D. 99

B. 97                    E. 100

C. 98

**Exercise 11:**    The median is

A. 96                    D. 99

B. 97                    E. 100

C. 98

**Exercise 12:**    The mode is

A. 96                    D. 99

B. 97                    E. 100

C. 98

**Let's look at the answers.**

**Answer 1:**   B: $-3 - 4(-5) - (-5)^2 = -3 + 20 - 25 = -8$.

**Answer 2:**   B: $0 - 0 - 2 = -2$.

**Answer 3:**   E: $95(4) = 380$ points; $90 + 91 + 98 = 279$ points. The fourth test would have to be $380 - 279 = 101\%$ (not possible).

**Answer 4:**   B: The median is 100; the mode is 100; for the mean, we can use 2 and 3 instead of 20 and 30 because the ratio is the same: $\dfrac{2(90) + 3(100)}{5} = 96$.

So the sum is $100 + 100 + 96 = 296$.

**Answer 5:**   D: $Mean = \dfrac{\text{total value of items}}{\text{total number of items}} = \dfrac{mx + ny + pz}{m + n + p}$.

**Answer 6:**   E: The number is not divisible by 11 because $3 - 2 = 1$, which is not a multiple of 11. It is not divisible by 3 because $3 + 2 = 5$ is not divisible by 3. Because is it not divisible by 11 or 3, it is not divisible by 33. Finally, it is not divisible by 10 because it does not end in a 0. So the answer is E because all numbers are divisible by 1.

**Answer 7:**   C: $\dfrac{8 + 10 + 10 + 16 + 16 + 18}{6} = 13$.

**Answer 8:**   C: There are an even number of numbers, so we have to take the average of the middle two: $\dfrac{10 + 16}{2} = 13$.

**Answer 9:**   E: It's bimodal; the modes are 10 and 16, each appearing twice.

**Answer 10:**   B: The mean is the longest measure to compute:
$\dfrac{4 \times 100 + 3 \times 98 + 2 \times 95 + 86}{10} = 97$.

**Answer 11:**   C: The median is determined by putting all of the numbers in order, so we have 100, 100, 100, 100, 98, 98, 98, 95, 95, 86. The middle terms are 98 and 98, so the median is 98.

**Answer 12:**   E: The mode is 100 because that is the most common score; there are four of them.

## Chapter 1 Quiz

1. List the even integers between $-4$ and 7.
2. List the multiples of seven between 42 and 73 inclusive.
3. List the primes between 100 and 110.
4. Write all the factors of 24.
5. Write 90 as the product of prime factors.
6. Why is $4.56 \times 10^{1000}$ divisible by 3 and not by 9?
7. $9 - 3^3 - 1^4 =$
8. $7 - 3(-5) - 7(2) =$
9. $-1^{10} + (-1)^{11} =$
10. $\dfrac{(-2)^3 - (-2)^4}{8 - (-4)^2} =$

For questions 11 and 12, let $a = -5$ and $d = 3$; evaluate the following:

11. $ad - a^2$
12. $ad^2 - (ad)^2$

For questions 13–16, consider the set $\{1, 2, 2, 5, 6, 8, 10\}$.

13. Find the mean.
14. Find the mode.
15. Find the median.
16. Find the range.

For questions 17–20, consider the set $\{-9, -9, -9, 4, 4, 4\}$.

17. Find the mean.
18. Find the mode.
19. Find the median.
20. Find the range.

## Answers to Chapter 1 Quiz

1. −2, 0, 2, 4, and 6; "between" doesn't include the end numbers. Remember, 0 is an even integer.

2. 42, 49, 56, 63, and 70; "inclusive" means to include the end number.

3. 101, 103, 107, and 109 (101 and 103, 107 and 109 are sometimes called twin primes because they differ by 2; see if you can find more of them).

4. 1, 2, 3, 4, 6, 8, 12, and 24.

5. $2 \times 3 \times 3 \times 5$, or $2 \times 3^2 \times 5$.

6. $4 + 5 + 6 = 15$ is divisible by 3 but not by 9.

7. $9 - 27 - 1 = -19$.

8. $7 + 15 - 14 = 8$.

9. $-1 + (-1) = -2$.

10. $\dfrac{(-8 - 16)}{(8 - 16)} = \dfrac{-24}{-8} = 3$.

11. $(-5)(3) - (-5)^2 = -15 - 25 = -40$.

12. $(-5)(3)^2 - [(-5)(3)]^2 = (-5)(9) - (-15)^2 = -45 - 225 = -270$.

13. $\dfrac{34}{7} = 4\dfrac{6}{7}$.

14. 2.

15. 5.

16. $10 - 1 = 9$.

17. $-\dfrac{15}{6} = -2\dfrac{1}{2}$.

18. −9 and 4 are both modes (bimodal).

19. $\dfrac{(-9 + 4)}{2} = -2\dfrac{1}{2}$.

20. $4 - (-9) = 13$.

Let's do some decimals, fractions, and percentages.

**Answer to "Bob Asks":** 9 cuts.

*"We are building a 60-foot fence in front of my house. A fence post is needed every 6 feet. How many fence posts are needed?"*

**Although** the COMPASS does allow calculators, it is necessary to be able to do some of the work without the calculator, especially fractions. Let's start with decimals.

## DECIMALS

*Rule 1:* When adding or subtracting, line up the decimal points.

**Example 1:**   Add: 3.14 + 234.7 + 86

**Solution:**
```
     3.14
   234.7
+  86.
─────────
   323.84
```

**Example 2:**   Subtract: 56.7 − 8.82

**Solution:**
```
   56.70
−   8.82
─────────
   47.88
```

*Rule 2:* In multiplying numbers, count the number of decimal places and add them. In the product, this will be the number of decimal places for the decimal point.

**Example 3:**   Multiply 45.67 by .987.

**Solution:**    The answer will be 45.07629. You will need to know the answer has five decimal places.

**Example 4:**    Multiply 2.8 by .6:

**Solution:**    The answer is 1.68.

**Example 5:**    What is the value of $2b^2 - 2.4b - 1.7$ if $b = .7$?

**Solution:**    The standard way to do this type of problem is to directly substitute.

$2(.7)^2 - 2.4(.7) - 1.7 = 2(.49) - 1.68 - 1.7 = .98 - 1.68 - 1.7 = -2.40$.

You will be given answers from which to choose, so approximations may work just as well and are quicker. $(.7)(.7) = .49$; times 2 is .98, or approximately 1. $-2.4$ times $.7 = 1.68$ or approximately 1.7; $1 - 1.7 = -.7$; added to $-1.7$ is $-2.4$. In this case, we get the same numerical value. If we didn't, our approximate answer would still probably be closer to the correct answer than to the other choices.

In the real world, and on this test, a good skill to save time is the ability to make reasonable approximations. This book will point out problems that can be approximated.

**Rule 3:**  When you divide, move the decimal point in the divisor and the dividend the same number of places.

**Example 6:**    Divide 23.1 by .004.

**Solution:**    In our heads, we write the problem as $\dfrac{23.1}{.004}$. We multiply the numerator

and denominator by 1,000 to move the decimal points the same number

of places to get whole numbers. We get $\dfrac{23.1}{.004} = \dfrac{23.1 \times 1,000}{.004 \times 1,000} = \dfrac{23,100}{4} =$

5,775. Note that when we multiply by 1, the fraction doesn't change, so for

$\dfrac{1,000}{1,000} = 1$, the fraction is the same.

**Rule 4:**  When reading a number with a decimal, read the whole part, only say the word *and* when you reach the decimal point, then read the part after the decimal point as if it were a whole number, and say the last decimal place. Whew!

**Example 7:**

| Number | Read |
|---|---|
| 4.3 | Four and three tenths |
| 2,006.73 | Two-thousand six and seventy-three hundredths |
| 1,000,017.009 | One million seventeen and nine thousandths |

**Let's do a couple of multiple-choice exercises.**

**Exercise 1:**    Which is smallest?

A. .04                D. .04444

B. .0401              E. .041

C. .04001

**Exercise 2:**    $\dfrac{7}{100} + \dfrac{8}{10,000} + \dfrac{9}{1,000,000} =$

A. 789               D. .070809

B. .0789             E. 078009

C. .07089

 **Let's look at the answers.**

**Answer 1:**    **A:** All have the same tenths (0) and hundredths (4). The rest of the places of A are zeros, so it's the smallest.

**Answer 2:**    **D:** Hundredths is the second decimal place, ten-thousandths is the fourth, and millionths is the sixth place.

Now, let's go over fractions.

# FRACTIONS

The top of a fraction is called the **numerator**; the bottom is the **denominator**.

*Rule 1:* If the bottoms of two fractions are the same, the bigger the top, the bigger the fraction.

**Example 8:**    Suppose I am a smart first grader. Can you explain to me which is bigger, $\dfrac{3}{5}$ or $\dfrac{4}{5}$?

**Solution:**    Suppose we have a pizza pie. Then $\frac{3}{5}$ means we divide a pie into 5 equal parts, and I get 3. And $\frac{4}{5}$ means I get 4 pieces out of 5. So $\frac{3}{5} < \frac{4}{5}$.

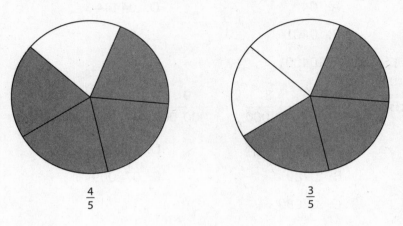

$$\frac{4}{5} \qquad\qquad\qquad \frac{3}{5}$$

*Rule 2:* If the tops of two fractions are the same, the bigger the bottom, the smaller the fraction.

**Example 9:**    Which fraction, $\frac{3}{5}$ or $\frac{3}{4}$, is bigger?

**Solution:**    Use another pizza pie example. In comparing $\frac{3}{5}$ and $\frac{3}{4}$, we get the same number of pieces (3). However, if the pie is divided into 4, the pieces are bigger, so $\frac{3}{5} < \frac{3}{4}$.

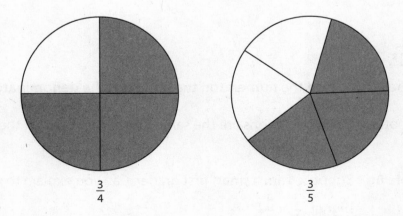

$$\frac{3}{4} \qquad\qquad\qquad \frac{3}{5}$$

*Rule 3:* If the tops and bottoms are different, find the least common denominator (LCD) and compare the tops.

Before we get into this section, the teacher in me (and maybe the purist in you) must tell you we really are talking about rational numbers, not fractions. There are two definitions.

**Definition 1:** A **rational** number is any integer divided by an integer, with the denominator not equaling zero.

**Definition 2:** A **rational** number is any repeating or terminating decimal.

 *Technically, $\frac{\pi}{6}$ is a fraction but not a rational number. We will use the term* fraction *here instead of rational number. If it is negative, we will say "negative fraction."*

Note the following facts about fractions:

- $3 < 4$, but $-3 > -4$. Similarly, $\frac{3}{5} < \frac{4}{5}$, but $-\frac{3}{5} > -\frac{4}{5}$. We will do more of this later.
- A fraction (positive) is bigger than 1 if the numerator is bigger than the denominator.
- A fraction is less than $\frac{1}{2}$ if the denominator is more than twice the numerator.
- To double a fraction, either multiply the numerator by two or divide the denominator by 2.
- Adding the same positive number to the numerator and denominator makes the fraction closer to one.
- Two fractions are **equivalent** if they can be reduced to the same fraction. $\frac{6}{9}$ and $\frac{10}{15}$ are equivalent because they both reduce to $\frac{2}{3}$.

**Q** **Let's do a few more multiple-choice exercises.**

**Exercise 3:**   Which fraction is largest?

A. $\frac{7}{4}$                     D. $\frac{12}{5}$

B. $\frac{11}{8}$                     E. $\frac{3}{100}$

C. $\frac{3}{2}$

**Exercise 4:**   Which fraction is smallest?

A. $\frac{3}{5}$                     D. $\frac{9}{16}$

B. $\frac{5}{11}$                     E. $\frac{100}{199}$

C. $\frac{7}{13}$

**Exercise 5:**    Which fraction is largest?

A. $\dfrac{1}{3}$                    D. $\dfrac{3}{.1}$

B. $\dfrac{1}{.3}$                    E. $\dfrac{3}{(.1)^2}$

C. $\dfrac{1}{(.3)^2}$

**Exercise 6:**    If $100 \le x \le 10{,}000$ and $.0001 \le y \le .01$, the smallest possible value of $\dfrac{x}{y}$ is

A. 10,000                    D. 10,000,000

B. 100,000                    E. 1,000,000,000

C. 1,000,000

**Exercise 7:**    Suppose we have $\dfrac{x+y}{x-y}$, where $8 \le x \le 10$ and $2 \le y \le 4$. The maximum possible value for this fraction is

A. $\dfrac{3}{2}$                    D. 3

B. $\dfrac{5}{3}$                    E. 6

C. $\dfrac{7}{3}$

 **Let's look at the answers.**

**Answer 3:**    D: It is the only fraction for which the top is more than double the bottom, so it is the only fraction with a value greater than 2.

**Answer 4:**    B: It is the only fraction less than one-half; the bottom is more than double the top.

**Answer 5:**    E: Answer choice A is less than 1. Comparing the other answer choices, C is bigger than B, and E is bigger than D. Because E has a bigger top and smaller bottom than C, it is larger. We will explore this in greater detail in the next chapter.

**Answer 6:**    A: To make a fraction as small as possible, we need a fraction with the smallest top ($x = 100$) and the biggest bottom ($y = .01$), or $\dfrac{100}{.01} = 10{,}000$.

**Answer 7:**    **D:** This one is not so simple. In this case, the extreme values, minimum (min) and maximum (max), occur at the "ends." We must try $x = 8$ and $10$, $y = 2$ and $4$, and all combinations of these numbers. In this case, the max occurs when $x = 8$ and $y = 4$.

## Adding and Subtracting Fractions

If the denominators are the same, add or subtract the tops, keep the bottom the same, and reduce if necessary.

$$\frac{7}{43} + \frac{11}{43} - \frac{2}{43} = \frac{16}{43}$$

$$\frac{2}{9} + \frac{4}{9} = \frac{6}{9} = \frac{2}{3}$$

$$\frac{a}{m} + \frac{b}{m} - \frac{c}{m} = \frac{a + b - c}{m}$$

There is much more to talk about if the denominators are unlike.

The quickest way to add (or subtract) fractions with different denominators, especially if they contain letters or the denominators are small (but different), is to multiply the top and bottom of each fraction by the least common multiple, LCM. This consists of three words: multiple, common, and least.

**Example 10:**   What is the LCM of 6 and 8?

**Solution:**   **Multiples** of 6 are 6, 12, 18, 24, 30, 36, 42, 48, 54, 60, 66, 72, 78, . . .

**Multiples** of 8 are 8, 16, 24, 32, 40, 48, 56, 64, 72, 80, . . .

**Common multiples** of 6 and 8 are 24, 48, 72, 96, 120, . . .

The **least common multiple** of 6 and 8 is 24.

When adding or subtracting fractions, multiply the top and bottom of each fraction by the LCM divided by the denominator of that fraction:

$$\frac{a}{b} - \frac{x}{y} = \left(\frac{a}{b} \times \frac{y}{y}\right) - \left(\frac{x}{y} \times \frac{b}{b}\right) = \frac{ay}{by} - \frac{bx}{by} = \frac{ay - bx}{by}$$

$$\frac{7}{20} - \frac{3}{11} = \left(\frac{7}{20} \times \frac{11}{11}\right) - \left(\frac{3}{11} \times \frac{20}{20}\right) = \frac{7(11) - 3(20)}{20(11)} = \frac{17}{220}$$

On the COMPASS, you must be able to perform these calculations quickly.

**Example 11:** Find the sums:

| Problem | Solution |
|---|---|
| a. $\dfrac{3}{8} + \dfrac{5}{6} =$ | $\dfrac{9}{24} + \dfrac{20}{24} = \dfrac{29}{24}$ |
| b. $\dfrac{3}{4} + \dfrac{5}{6} =$ | $\dfrac{3}{4} + \dfrac{5}{6} = \dfrac{9}{12} + \dfrac{10}{12} = \dfrac{19}{12}$ |
| c. $4\dfrac{3}{8} + 5\dfrac{5}{6} =$ | $4\dfrac{9}{24} + 5\dfrac{20}{24} = 9\dfrac{29}{24} = 9 + 1\dfrac{5}{24} = 10\dfrac{5}{24}$ |

Doing this problem on the COMPASS can take a long time by calculator. It is to your advantage to be able to do problems like this with paper and pencil.

## Multiplication of Fractions

To multiply fractions, multiply the numerators and multiply the denominators, reducing as you go. With multiplication, it is *not* necessary to have the same denominators.

$$\frac{3}{7} \times \frac{4}{11} = \frac{12}{77}$$

$$\frac{a}{b} \times \frac{c}{d} = \frac{a \times c}{b \times d}$$

## Division of Fractions

To divide fractions, invert the second fraction and multiply, reducing if necessary. To **invert** a fraction means to turn it upside down. The new fraction is called the **reciprocal** of the original fraction. So the reciprocal of $\dfrac{2}{3}$ is $\dfrac{3}{2}$; the reciprocal of $-5$ is $-\dfrac{1}{5}$; and the reciprocal of $a$ is $\dfrac{1}{a}$. if $a \neq 0$.

**Example 12:** Do the following calculations:

| Problem | Solution |
|---|---|
| a. $\dfrac{3}{4} \div \dfrac{11}{5} =$ | $\dfrac{3}{4} \times \dfrac{5}{11} = \dfrac{15}{44}$ |
| b. $\dfrac{m}{n} \div \dfrac{p}{q} =$ | $\dfrac{m}{n} \times \dfrac{q}{p} = \dfrac{m \times q}{n \times p}$ |

c. $\dfrac{1}{4} \div 5 =$     $\dfrac{1}{4} \times \dfrac{1}{5} = \dfrac{1}{20}$

d. $3\dfrac{1}{5} \times 1\dfrac{1}{6} =$     $\dfrac{16}{5} \times \dfrac{7}{6} = \dfrac{8}{5} \times \dfrac{7}{3} = \dfrac{56}{15}$ or $3\dfrac{11}{15}$     ✓

Let's do a few problems.

**Example 13:** Problem     Solution

a. $\dfrac{7}{9} - \dfrac{3}{22} =$     $\dfrac{127}{198}$

b. $\dfrac{3}{4} + \dfrac{5}{6} - \dfrac{1}{8} =$     $\dfrac{35}{24}$, or $1\dfrac{11}{24}$

c. $\dfrac{3}{10} + \dfrac{2}{15} - \dfrac{4}{5} =$     $\dfrac{-11}{30}$

d. $\dfrac{1}{4} + \dfrac{1}{8} + \dfrac{7}{16} =$     $\dfrac{13}{16}$

e. $2 + \dfrac{2}{3} + \dfrac{2}{9} + \dfrac{2}{27} =$     $2\dfrac{26}{27}$, or $\dfrac{80}{27}$

f. $\dfrac{5}{24} - \dfrac{7}{18} =$     $-\dfrac{13}{72}$

g. $\dfrac{10}{99} - \dfrac{9}{100} =$     $\dfrac{109}{9,900}$

h. $\dfrac{7}{9} \times \dfrac{5}{3} =$     $\dfrac{35}{27}$, or $1\dfrac{8}{27}$

i. $\dfrac{11}{12} \div \dfrac{9}{11} =$     $\dfrac{121}{108}$, or $1\dfrac{13}{108}$

j. $\dfrac{5}{9} \times \dfrac{6}{7} =$     $\dfrac{10}{21}$

k. $\dfrac{12}{13} \div \dfrac{8}{39} =$     $\dfrac{9}{2}$, or $4\dfrac{1}{2}$

l. $\dfrac{10}{12} \div \dfrac{15}{40} =$     $\dfrac{20}{9}$, or $2\dfrac{2}{9}$

m. $\dfrac{2}{3} \div 12 =$                                    $\dfrac{1}{18}$

n. $\dfrac{2}{3} \times \dfrac{3}{4} \times \dfrac{4}{5} \times \dfrac{5}{6} \times \dfrac{6}{7} =$          $\dfrac{2}{7}$

o. $\dfrac{5}{8} \times \dfrac{7}{6} \div \dfrac{35}{24} =$                          $\dfrac{1}{2}$

p. $3\dfrac{2}{3} + 4\dfrac{3}{4} =$                            $8\dfrac{5}{12}$

q. $5\dfrac{1}{6} - 3\dfrac{5}{6} =$                            $1\dfrac{1}{3}$

r. $7\dfrac{3}{7} - 2\dfrac{1}{2} =$                            $4\dfrac{13}{14}$

s. $3\dfrac{4}{5} \times \dfrac{3}{38} =$                          $\dfrac{3}{10}$

t. $4\dfrac{1}{5} \div 8\dfrac{2}{5} =$                          $\dfrac{1}{2}$

**Q** **Let's do a couple more multiple-choice exercises.**

**Exercise 8:**   What number, when multiplied by $\dfrac{3}{4}$, gives $\dfrac{7}{8}$?

A. $\dfrac{21}{32}$            D. $\dfrac{8}{7}$

B. $\dfrac{6}{5}$            E. $\dfrac{4}{3}$

C. $\dfrac{7}{6}$

**Exercise 9:**   The average (mean) of $\dfrac{1}{4}$ and $\dfrac{1}{8}$ is

A. $\dfrac{1}{12}$            D. $\dfrac{5}{32}$

B. $\dfrac{1}{6}$            E. $\dfrac{7}{24}$

C. $\dfrac{3}{16}$

**Let's look at the answers.**

**Answer 8:**   **C:** We need to solve $\frac{3}{4}x = \frac{7}{8}$. Then $x = \frac{7}{8} \times \frac{4}{3} = \frac{7}{6}$. You may need to be able to do this is your head.

**Answer 9:**   **C:** $\frac{\left(\frac{1}{4} + \frac{1}{8}\right)}{2} = \frac{\left(\frac{3}{8}\right)}{2} = \frac{3}{16}$.

## Changing from Decimals to Fractions and Back

To change from a decimal to a fraction, we read it and write it.

**Example 14:**   Change 4.37 to a fraction.

**Solution:**   We read it as 4 and 37 hundredths: $4\frac{37}{100} = \frac{437}{100}$, if necessary. That's it.

**Example 15:**   Change to decimals:

   **a.** $\frac{7}{4}$                **b.** $\frac{1}{6}$

**Solution:**   For such fractions, the decimal will either terminate or repeat.

   **a.** Divide 4 into 7.0000: $7.0000 \div 4 = 1.75$

   **b.** Divide 6 into 1.0000: $1.0000 \div 6 = .1666\ldots = .1\overline{6}$

The bar over the 6 means it repeats forever; for example, $.3454545\ldots = .3\overline{45}$. The bar over the 45 means 45 repeats forever, but not the 3.

## PERCENTAGES

% means hundredths: $1\% = \frac{1}{100} = .01$.

Follow these rules to change between percentages and decimals and fractions:

*Rule 1:* To change a percentage to a decimal, move the decimal point two places to the left and drop the % sign.

*Rule 2:* To change a decimal to a percentage, move the decimal point two places to the right and add a % sign.

*Rule 3:* To change from a percentage to a fraction, divide by 100% and simplify, or change the % sign to $\frac{1}{100}$ and multiply.

*Rule 4:* To change a fraction to a percentage, first change to a decimal, and then to a percentage.

**Example 16:** Change 12%, 4%, and .7% to decimals.

**Solutions:**    12% = 12.% = .12; 4% = 4.% = .04; .7% = .007.

**Example 17:** Change .734, .2, and 34 to percentages.

**Solutions:**    .734 = 73.4%; .2 = 20%; 34 = 34. = 3400%.

**Example 18:** Change 42% to a fraction.

**Solution:**    $42\% = .42 = \dfrac{42}{100} = \dfrac{21}{50}$, or $42\% = 42 \times \dfrac{1}{100} = \dfrac{42}{100} = \dfrac{21}{50}$

**Example 19:** Change $\dfrac{7}{4}$ to a percentage.

**Solution:**    $\dfrac{7}{4} = 1.75 = 175\%$

It is important to know these for the test.

| Fraction | Decimal | Percentage | Fraction | Decimal | Percentage |
|---|---|---|---|---|---|
| $\frac{1}{2}$ | .5 | 50% | $\frac{1}{8}$ | .125 | $12\frac{1}{2}\%$ |
| $\frac{1}{4}$ | .25 | 25% | $\frac{3}{8}$ | .375 | $37\frac{1}{2}\%$ |
| $\frac{3}{4}$ | .75 | 75% | $\frac{5}{8}$ | .625 | $62\frac{1}{2}\%$ |
| $\frac{1}{5}$ | .2 | 20% | $\frac{7}{8}$ | .875 | $87\frac{1}{2}\%$ |
| $\frac{2}{5}$ | .4 | 40% | $\frac{1}{16}$ | .0625 | $6\frac{1}{4}\%$ |
| $\frac{3}{5}$ | .6 | 60% | $\frac{3}{16}$ | .1875 | $18\frac{3}{4}\%$ |
| $\frac{4}{5}$ | .8 | 80% | $\frac{5}{16}$ | .3125 | $31\frac{1}{4}\%$ |
| $\frac{1}{6}$ | .1$\overline{6}$ | $16\frac{2}{3}\%$ | $\frac{7}{16}$ | .4375 | $43\frac{3}{4}\%$ |
| $\frac{1}{3}$ | .$\overline{3}$ | $33\frac{1}{3}\%$ | $\frac{9}{16}$ | .5625 | $56\frac{1}{4}\%$ |
| $\frac{2}{3}$ | .$\overline{6}$ | $66\frac{2}{3}\%$ | $\frac{11}{16}$ | .6875 | $68\frac{3}{4}\%$ |
| $\frac{5}{6}$ | .8$\overline{3}$ | $83\frac{1}{3}\%$ | $\frac{13}{16}$ | .8125 | $81\frac{1}{4}\%$ |
| | | | $\frac{15}{16}$ | .9375 | $93\frac{3}{4}\%$ |

If you can't memorize all of these with denominator 16, memorize at least $\dfrac{1}{16}$ and $\dfrac{15}{16}$.

If you are good at doing percentage problems, skip this next section. Otherwise, here's a really easy way to do percentage problems. Make the following pyramid:

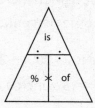

**Example 20:**  What is 12% of 1.3?

**Solution:**

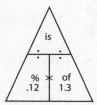

Put .12 in the % box (always change to a decimal in this box) and 1.3 in the "of" box. It tells us to multiply .12 × 1.3 = .156. That's all there is to it.

**Example 21:**  8% of what is 32?

**Solution:**

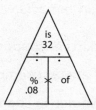

.08 goes in the % box. 32 goes in the "is" box. 32 ÷ .08 = 400.

**Example 22:**  9 is what % of 8?

**Solution:**

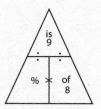

9 goes in the "is" box. 8 goes in the "of" box. 9 ÷ 8 × 100% = 112.5%.

The goal is to be able to do percentage problems without writing the pyramid.

**Example 23:** In ten years, the population increases from 20,000 to 23,000. Find the actual increase and the percentage increase.

**Solution:** The actual increase is $23,000 - 20,000 = 3,000$. The percentage increase is $\dfrac{3000}{20,000} \times 100\% = 15\%$ increase.

**Example 24:** The cost of producing widgets decreased from 60 cents to 50 cents. Find the actual decrease and percentage decrease.

**Solution:** $60 - 50 = 10$ cent actual decrease; $\dfrac{10}{60} = 16\dfrac{2}{3}\%$ decrease.

**Note** *Percentage increases and decreases are figured on the original amount.*

**Example 25:** The cost of a $2,000 large-screen TV set is decreased by 30%. If there is a 7% sales tax, how much does it cost?

**Solution:** $2,000 \times .30 = $600 discount. $2,000 - $600 = $1,400 cost. $1,400 \times .07 is $98. The total price is $1,400 + $98 = $1,498.

**Note** *If we took 70% (or 100% − 30%) of $2,000, we would immediately get the cost.*

There is an interesting story about why women wear miniskirts in London, England. It seems that the sales tax is $12\dfrac{1}{2}\%$ on clothes! But children's clothes are tax exempt. A girl's dress is any dress where the skirt is less than 24 inches; so that is why women in London wear miniskirts!

**Ⓠ    Let's do some multiple-choice exercises.**

**Exercise 10:** The product of 2 and $\dfrac{1}{89}$ is:

A. $2\dfrac{1}{89}$          D. $\dfrac{2}{89}$

B. $1\dfrac{88}{89}$          E. $\dfrac{1}{172}$

C. 172

**Exercise 11:** $\dfrac{1}{50}$ of 2% of .02 is

    **A.** .08             **D.** .000008

    **B.** .008            **E.** .00000008

    **C.** .0008

**Exercise 12:** 30% of 20% of a number is the same as 40% of what percentage of the same number?

    **A.** 10             **D.** 18

    **B.** $12\dfrac{1}{2}$         **E.** Can't be determined without the number

    **C.** 15

**Exercise 13:** A fraction between $\dfrac{3}{43}$ and $\dfrac{4}{43}$ is

    **A.** $\dfrac{1}{9}$           **D.** $\dfrac{7}{86}$

    **B.** $\dfrac{3}{28}$          **E.** $\dfrac{9}{1849}$

    **C.** $\dfrac{5}{47}$

**Exercise 14:** The reciprocal of $2 - \dfrac{3}{4}$ is

    **A.** $\dfrac{1}{2} - \dfrac{4}{3}$       **D.** $\dfrac{5}{4}$

    **B.** $-\dfrac{5}{4}$         **E.** $\dfrac{4}{5}$

    **C.** $-\dfrac{4}{5}$

**Exercise 15:** A price increase of 20% followed by a decrease of 20% means the price is

    **A.** Up 4%         **D.** Down 2%

    **B.** Up 2%         **E.** Down 4%

    **C.** The original price

**Exercise 16:**   A price decreases 20% followed by a 20% increase. The final price is

    **A.** Up 4%                        **D.** Down 2%

    **B.** Up 2%                        **E.** Down 4%

    **C.** The original price

**Exercise 17:**   A 50% discount followed by another 50% discount gives a discount of

    **A.** 25%                        **D.** 90%

    **B.** 50%                        **E.** 100%

    **C.** 75%

**Exercise 18:**   $83\frac{1}{3}\%$ of $37\frac{1}{2}\%$ is

    **A.** $\frac{1}{2}$                        **D.** $\frac{5}{16}$

    **B.** $\frac{2}{5}$                        **E.** $\frac{3}{32}$

    **C.** $\frac{1}{4}$

**Exercise 19:**   Which is equivalent to .0625?

    **A.** $\frac{3}{8}$                        **D.** $\frac{1}{18}$

    **B.** $\frac{1}{8}$                        **E.** $\frac{1}{80}$

    **C.** $\frac{1}{16}$

**Exercise 20:**   If the fractions $\frac{19}{24}, \frac{1}{2}, \frac{3}{8}, \frac{3}{4},$ and $\frac{9}{16}$ were ordered least to greatest, the middle number would be

    **A.** $\frac{19}{24}$                        **D.** $\frac{3}{4}$

    **B.** $\frac{1}{2}$                        **E.** $\frac{9}{16}$

    **C.** $\frac{3}{8}$

**Exercise 21:** $\dfrac{31}{125} =$

A. 0.320

D. 0.252

B. 0.310

E. 0.248

C. 0.288

**Ⓐ Let's look at the answers.**

**Answer 10:** D: $\dfrac{2}{1} \times \dfrac{1}{89} = \dfrac{2}{89}$.

**Answer 11:** D: $.02 \times .02 \times .02 = .000008$.

**Answer 12:** C: We can forget the number and forget the percentages. $(30)(20) = (40)(?)$. $? = 15$.

**Answer 13:** D: $\dfrac{3}{43} = \dfrac{6}{86}$; $\dfrac{4}{43} = \dfrac{8}{86}$; between is $\dfrac{7}{86}$. As another example like this one, to get nine numbers between $\dfrac{3}{43}$ and $\dfrac{4}{43}$, multiply both fractions, top and bottom, by 10 (one more than nine), and the fractions in between would be $\dfrac{31}{430}, \dfrac{32}{430}, \dfrac{33}{430}$, etc.

**Answer 14:** E: $2 - \dfrac{3}{4} = \dfrac{5}{4}$. Its reciprocal is $\dfrac{4}{5}$.

**Answer 15:** E: Let $100 be the original price. A 20% increase means a $20 increase. Then a 20% decrease is ($120) $\times$ (.20) = $24. $120 − $24 = $96, or a $4 decrease from the original price. $\dfrac{\$4}{\$100}$ is a 4% decrease.

**Answer 16:** E: Suppose we have $100. Increased by 20%, we have $120. But another 20% less (now on a larger amount) is $24 less. We are at $96, down 4% from the original price. Suppose we take 20% off first. We are at $80. Now 20% up (on a smaller amount) is $16. We are again at $96 or again down 4% from the original price.

**Answer 17:    C:** $100 discounted 50% is $50. 50% of $50 is $25, for a 75% discount. This is why after December, a 50% discount followed by 20% is 60%, but sounds like 70% off.

**Answer 18:    D:** If you memorized the fraction table, you would recognize this problem as the same as $\left(\dfrac{5}{6}\right)\left(\dfrac{3}{8}\right) = \dfrac{5}{16}$.

**Answer 19:    C:** It is easy if you've memorized the fraction table.

**Answer 20:    E:** If we changed each number to a fraction with a denominator of 48 (the LCM), we would get $\dfrac{19}{24} = \dfrac{38}{48}, \dfrac{1}{2} = \dfrac{24}{48}, \dfrac{3}{8} = \dfrac{18}{48}, \dfrac{3}{4} = \dfrac{36}{48}$, and $\dfrac{9}{16} = \dfrac{27}{48}$, and it is easy to see that $\dfrac{27}{48} = \dfrac{9}{16}$ is the middle number.

**Answer 21:    E:** There are two ways to do this problem. If you recognize that $\dfrac{31}{125} < \dfrac{1}{4} (= .25)$ because the bottom is more than 4 times the top, you see that answer choice E is the only one less than $\dfrac{1}{4}$. Another way to answer this question is to multiply the top and bottom of $\dfrac{31}{125}$ by 8, and get $\dfrac{248}{1000} = .248$, which is answer choice E.

## Chapter 2 Quiz

1. $51 + 5.1 + .51 =$

2. $23.4 - 3.49 =$

3. $(.12)(.012) =$

4. $\dfrac{.12}{.012} =$

5. Change $\dfrac{9}{8}$ to a decimal and a percent.

6. Change .5% to a decimal and a fraction.

7. Change 3.7 to a fraction and a percent.

8. 14% of 1.4 is what?

9. 12 is what percent of 9?

10. 14% of what is 6.3?

11. Reduce $\dfrac{26}{65}$.

12. Write $\dfrac{100}{46}$ as a reduced mixed number.

13. $2\dfrac{5}{8} + 4\dfrac{5}{6} =$

14. $12 - 2\dfrac{3}{7} =$

15. $3\dfrac{1}{2} \times 4\dfrac{1}{2} =$

16. $5\dfrac{1}{4} \div 2\dfrac{5}{8} =$

17. Find the maximum value of $\dfrac{x}{y}$ if $x$ and $y$ must be from the set $\{4, 5, 6, 7\}$.

18. An $80 radio is discounted 20%. If the state sales tax is 5%, how much does it cost?

19. A stock decreases from $20 a share to $16 a share. Find the percentage decrease.

20. A shirt sold for $45, 75% of the original cost. Find the original cost.

## Answers to Chapter 2 Quiz

1. $51 = 51.0$; the answer is 56.61.

2. 19.91.

3. 0.00144.

4. 10.

5. 1.125; 112.5%.

6. $0.005; \dfrac{5}{1000} = \dfrac{1}{200}$.

7. $\dfrac{37}{10}$, or $3\dfrac{7}{10}$; 370%.

8. $0.14 \times 1.4 = 0.196$.

9. $\dfrac{12}{9} \times 100\% = 133\dfrac{1}{3}\%$.

10. $\dfrac{6.3}{.14} = 45$.

11. $\dfrac{2}{5}$.

12. $\dfrac{50}{23} = 2\dfrac{4}{23}$.

13. $2\dfrac{15}{24} + 4\dfrac{20}{24} = 6\dfrac{35}{24} = 6 + 1\dfrac{11}{24} = 7\dfrac{11}{24}$.

14. $11\dfrac{7}{7} - 2\dfrac{3}{7} = 9\dfrac{4}{7}$.

15. $\dfrac{7}{2} \times \dfrac{9}{2} = \dfrac{63}{4} = 15\dfrac{3}{4}$.

16. $\dfrac{21}{4} \times \dfrac{8}{21} = 2$.

17. $\dfrac{7}{4}$; select the largest top number and the smallest bottom number.

18. $100\% - 20\% = 80\%$; $.80 \times \$80 = \$64$; $\$64 + .05 \times \$64 = \$67.20$.

19. $20 - 16 = 4$; $\dfrac{4}{20} \times 100\% = 20\%$.

20. $\dfrac{45}{.75}$, or $\dfrac{45}{\left(\frac{3}{4}\right)} = 45 \times \dfrac{4}{3} = 15 \times 4 = \$60$.

Enough! Let's do some algebra now. We will see these topics throughout the book.

**Answer to "Bob Asks":** 11 fence posts; you need one in the beginning.

*A bell rings at even intervals forever. If it rings 6 times in 5 seconds, how many times will it ring in 30 seconds? (Hint: This is like the fence post question at the beginning of Chapter 2.)*

**Exponents** are a very popular topic. They are a good test of knowledge and thinking, are short to write, and it is relatively easy to make up new problems. Let's review some basic rules of exponents.

| <u>Rule</u> | <u>Examples</u> |
|---|---|
| 1. $x^m x^n = x^{m+n}$ | $x^6 x^4 x = x^{11}$ and $(x^6 y^7)(x^4 y^{10}) = x^{10}y^{17}$ |
| 2. $\dfrac{x^m}{x^n} = x^{m-n}$ or $\dfrac{1}{x^{n-m}}$ | $\dfrac{x^8}{x^6} = x^2, \dfrac{x^3}{x^7} = \dfrac{1}{x^4}$, and $\dfrac{x^4 y^5 z^9}{x^9 y^2 z^9} = \dfrac{y^3}{x^5}$ |
| 3. $(x^m)^n = x^{mn}$ | $(x^5)^7 = x^{35}$ |
| 4. $(xy)^n = x^n y^n$ | $(xy)^3 = x^3 y^3$ and $(x^7 y^3)^{10} = x^{70} y^{30}$ |
| 5. $\left(\dfrac{x}{y}\right)^n = \dfrac{x^n}{y^n}$ | $\left(\dfrac{x}{y}\right)^6 = \dfrac{x^6}{y^6}$ and $\left(\dfrac{y^4}{z^5}\right)^3 = \dfrac{y^{12}}{z^{15}}$ |
| 6. $x^{-n} = \dfrac{1}{x^n}$ and $\dfrac{1}{x^{-m}} = x^m$ | $2^{-3} = \dfrac{1}{2^3} = \dfrac{1}{8}, \dfrac{1}{4^{-3}} = 4^3 = 64,$ $\dfrac{x^{-4} y^{-5} z^6}{x^{-6} y^4 z^{-1}} = \dfrac{x^6 z^6 z^1}{x^4 y^5 y^4} = \dfrac{x^2 z^7}{y^9},$ and $\left(\dfrac{x^3}{y^{-4}}\right)^{-2} = \left(\dfrac{y^{-4}}{x^3}\right)^2 = \dfrac{y^{-8}}{x^6} = \dfrac{1}{x^6 y^8}$ |
| 7. $x^0 = 1, x \neq 0;$ $0^0$ is indeterminate | $(7ab)^0 = 1$ and $7x^0 = 7(1) = 7$ |

8. $x^{\frac{p}{r}} = x^{\frac{power}{root}}$

Even though either order of computing the power and root gives the same answer, we usually do the root first. $25^{\frac{3}{2}} = \left(\sqrt{25}\right)^3 = 5^3 = 125$;

$$4^{-\frac{3}{2}} = \frac{1}{4^{\frac{3}{2}}} = \frac{1}{\left(\sqrt{4}\right)^3} = \left(\frac{1}{2}\right)^3 = \frac{1}{8}$$

It will save time if you know the following:

| | | | | | |
|---|---|---|---|---|---|
| $2^2 = 4$ | $2^3 = 8$ | $2^4 = 16$ | $2^5 = 32$ | $2^6 = 64$ | $2^7 = 128$ |
| $2^8 = 256$ | $2^9 = 512$ | $2^{10} = 1024$ | $3^2 = 9$ | $3^3 = 27$ | $3^4 = 81$ |
| $3^5 = 243$ | $3^6 = 729$ | $4^2 = 16$ | $4^3 = 64$ | $4^4 = 256$ | $5^2 = 25$ |
| $5^3 = 125$ | $6^2 = 36$ | $6^3 = 216$ | $7^2 = 49$ | $8^2 = 64$ | $9^2 = 81$ |
| $10^2 = 100$ | $11^2 = 121$ | $12^2 = 144$ | $13^2 = 169$ | $14^2 = 196$ | $15^2 = 225$ |
| $16^2 = 256$ | $17^2 = 289$ | $18^2 = 324$ | $19^2 = 361$ | $20^2 = 400$ | $21^2 = 441$ |
| $22^2 = 484$ | $23^2 = 529$ | $24^2 = 576$ | $25^2 = 625$ | $26^2 = 676$ | $27^2 = 729$ |
| $28^2 = 784$ | $29^2 = 841$ | $30^2 = 900$ | $31^2 = 961$ | $32^2 = 1,024$ | |

Over the years, I have asked my students to memorize these powers for a number of reasons. In math, these powers occur often. They occur when we use the Pythagorean theorem (mostly the squares). They are used going backward to find roots (next chapter). Also, they show number patterns of the squares (look at the last digits of each of the squares).

Even though the COMPASS does not ask comparison-type questions, it is necessary to be able to compare $x$ and $x^2$. Similar results hold for $x^3$, $x^4$, and so on.

If $x > 1$, then $x^2 > x$, because $4^2 > 4$.

If $x = 1$, then $x^2 = x$, because $1^2 = 1$.

If $0 < x < 1$, then $x > x^2$, because $\frac{1}{2} > \left(\frac{1}{2}\right)^2 = \frac{1}{4}$!!

If $x = 0$, then $x = x^2$, because $0 = 0^2$.

If $x < 0$, then $x^2 > x$, because the square of a negative is a positive.

Also recall that if $0 < x < 1$, then $\dfrac{1}{x} > 1$, and if $x > 1$, then $0 < \dfrac{1}{x} < 1$.

Also if $-1 < x < 0$, then $\dfrac{1}{x} < -1$, and if $x < -1$, then $-1 < \dfrac{1}{x} < 0$.

Here are some exponential problems to practice, including some with negative exponents.

**Example 1:**  Simplify the following:

<u>Problem</u>                                                       <u>Solution</u>

**a.** $(-3a^4bc^6)\,(-5ab^7c^{10})\,(-100a^{100}b^{200}c^{2000}) =$     $-1{,}500a^{105}b^{208}c^{2016}$

**b.** $(10ab^4c^7)^3 =$     $1{,}000a^3b^{12}c^{21}$

**c.** $(4x^6)^2\,(10x)^3 =$     $16{,}000x^{15}$

**d.** $((2b^4)^3)^2 =$     $64b^{24}$

**e.** $(-b^6)^{101} =$     $-b^{606}$

**f.** $(-ab^8)^{202} =$     $a^{202}b^{1616}$

**g.** $\dfrac{24e^9f^7g^5}{72e^9f^{11}g^7} =$     $\dfrac{1}{3f^4g^2}$

**h.** $\dfrac{\left(x^4\right)^3}{x^4} =$     $x^8$

**i.** $\left(\dfrac{m^3n^4}{m^7n}\right)^5 =$     $\dfrac{n^{15}}{m^{20}}$

**j.** $\left(\dfrac{\left(p^4\right)^3}{\left(p^6\right)^5}\right)^{10} =$     $\dfrac{1}{p^{180}}$

**k.** $(-10a^{-4}b^5c^{-2})(4a^{-7}b^{-1}) =$     $\dfrac{-40b^4}{a^{11}c^2}$

**l.** $(3ab^{-3}c^4)^{-3} =$     $\dfrac{b^9}{27a^3c^{12}}$

m. $(3x^4)^{-4}\left(\left(\frac{1}{9x^8}\right)^{-1}\right)^2 =$     1

n. $(2x^{-4})^2(3x^{-3})^{-2} =$     $\dfrac{4}{9x^2}$

o. $\left(\dfrac{(2y^3)^{-2}}{(4x^{-5})}\right)^{-2} =$     $\dfrac{256y^{12}}{x^{10}}$

**Q** **Now, let's try some multiple-choice exercises.**

**Exercise 1:**   $0 < x < 1$: Arrange in order of smallest to largest: $x, x^2, x^3$.

A. $x < x^2 < x^3$      D. $x^3 < x^2 < x$

B. $x < x^3 < x^2$      E. $x^3 < x < x^2$

C. $x^2 < x < x^3$

**Exercise 2:**   $-1 < x < 0$: Arrange in order of largest to smallest: $x^2, x^3, x^4$.

A. $x^4 > x^3 > x^2$      D. $x^2 > x^3 > x^4$

B. $x^4 > x^2 > x^3$      E. $x^2 > x^4 > x^3$

C. $x^3 > x^4 > x^2$

**Exercise 3:**   $0 < x < 1$.

I:  $x > \dfrac{1}{x^2}$

II: $\dfrac{1}{x^2} > \dfrac{1}{x^4}$

III: $x - 1 > \dfrac{1}{x - 1}$

Which statement(s) are always true?

A. None      D. III only

B. I only      E. All

C. II only

**Exercise 4:**   $(5ab^3)^3 =$

A. $15ab^6$      D. $125a^3b^6$

B. $75ab^6$      E. $125a^3b^9$

C. $125ab^9$

**Exercise 5:** $\dfrac{\left(2x^5\right)^3\left(3x^{10}\right)^2}{6x^{15}} =$

A. 1

B. $x^{20}$

C. $2x^{20}$

D. $12x^{20}$

E. $12x^{210}$

**Exercise 6:** $\left(\dfrac{12x^6}{24x^9}\right)^3 =$

A. $1728x^{27}$

B. $\dfrac{1}{1728x^{27}}$

C. $\dfrac{1}{6x^9}$

D. $\dfrac{1}{8x^9}$

E. $\dfrac{1}{8x^{27}}$

**Exercise 7:** $\dfrac{\left(4x^4\right)^3}{\left(8x^6\right)^2} =$

A. $\dfrac{1}{2}$

B. 1

C. $\dfrac{1}{2}x^{28}$

D. $x^{28}$

E. $\dfrac{1}{2x}$

**Exercise 8:** $-1 \le x \le 5$. Where is $x^2$ located?

A. $-1 \le x^2 \le 5$

B. $0 \le x^2 \le 25$

C. $1 \le x^2 \le 5$

D. $1 \le x^2 \le 10$

E. $1 \le x^2 \le 25$

**Exercise 9:** $2^m + 2^m =$

A. $2^{m+1}$

B. $2^{m+2}$

C. $2^{m+4}$

D. $2^{m^2}$

E. $4^m$

**Exercise 10:** $\dfrac{m^{-5}n^6p^{-2}}{m^{-3}n^9p^0} =$

A. $m^2n^3p^2$

D. $\dfrac{1}{m^2n^3p^2}$

B. $\dfrac{1}{m^2n^3}$

E. None of these

C. $\dfrac{m^2}{n^3p^2}$

**Exercise 11:** If $8^{2n+1} = 2^{n+18}$; $n =$

A. 3

D. 13

B. 7

E. 17

C. 10

**Exercise 12:** $p = 4^n$; $4p =$

A. $4^{n+1}$

D. $16^p$

B. $4^{n+2}$

E. $64^p$

C. $3^{n+4}$

**Exercise 13:** $y^3 = 64$; $y^{-2} =$

A. $-4$

D. $\dfrac{1}{16}$

B. $-\dfrac{1}{8}$

E. $-\dfrac{1}{16}$

C. $\dfrac{1}{8}$

**Exercise 14:** Suppose $x^2 = y^2$.

I:  $x = y$.

II:  $x = -y$.

III:  $x^2 = xy$.

Which statements are always true?

A. None

D. III only

B. I only

E. All statements are true

C. II only

**Exercise 15:** $3^{-2} =$

A. $\dfrac{1}{3}$                  D. $-\dfrac{1}{9}$

B. $\dfrac{1}{6}$                  E. $-\dfrac{1}{6}$

C. $\dfrac{1}{9}$

**Exercise 16:** $x^{\frac{3}{4}} = 8; x =$

A. $\dfrac{32}{3}$             D. 256

B. 16                 E. 1024

C. 64

**Exercise 17:** $n$ is an integer, and $(-2)^{6n} = 8^{n+4}; n =$

A. 2                  D. 6

B. 3                  E. 8

C. 4

 **Let's look at the answers.**

**Answer 1:** D: If $0 < x < 1$, the higher the power, the smaller the number.

**Answer 2:** E: Take, for example, $x = -\dfrac{1}{2}. \left(-\dfrac{1}{2}\right)^2 = \dfrac{1}{4}; \left(-\dfrac{1}{2}\right)^3 = -\dfrac{1}{8};$ $\left(-\dfrac{1}{2}\right)^4 = \dfrac{1}{16}.$ Be careful! This exercise asks for largest to smallest. Notice that $x^3$ has to be the smallest because it is the only negative number, so the answer choices are reduced to two, B and E.

**Answer 3:** D: Statement I is false: $0 < x < 1$, so $\dfrac{1}{x} > 1$ and $\dfrac{1}{x^2} > 1$ also. Statement II is false: $x^2 > x^4$, so $\dfrac{1}{x^2} < \dfrac{1}{x^4}$. Statement III is true: $0 < x < 1$; so $-1 < x - 1 < 0$; this means $\dfrac{1}{x-1} < -1$

**Answer 4:** E: $5^3 a^3 (b^3)^3 = 125 a^3 b^9.$

**Answer 5:**    D: $\left(\dfrac{8 \times 9}{6}\right) x^{15 + 20 - 15} = 12x^{20}$.

**Answer 6:**    D: $\left(\dfrac{1}{2x^3}\right)^3 = \dfrac{1}{8x^9}$.

**Answer 7:**    B: The numerator and denominator of the fraction each equal $64x^{12}$.

**Answer 8:**    B: This is very tricky because 0 is between $-1$ and 5, and $0^2 = 0$.

**Answer 9:**    A: This is one of the few truly hard problems because it is an addition problem:

$$2^m + 2^m = 1 \times 2^m + 1 \times 2^m = 2 \times 2^m = 2^1 \times 2^m = 2^{m+1}.$$

Similarly, $3^m + 3^m + 3^m = 3^{m+1}$ and four $4^m$ terms added equal $4^{m+1}$.

**Answer 10:**    D: $\dfrac{m^{-5}n^6p^{-2}}{m^{-3}n^9p^0} = \dfrac{m^3n^6}{m^5n^9p^2} = \dfrac{1}{m^2n^3p^2}$.

**Answer 11:**    A: $8^{2n+1} = (2^3)^{2n+1} = 2^{n+18}$. If the bases are equal, the exponents must be equal. $3(2n+1) = n + 18$; so $n = 3$.

**Answer 12:**    A: $p = 4^n$; $4p = 4(4^n) = 4^1 4^n = 4^{n+1}$.

**Answer 13:**    D: $y^3 = 64$; $y = 4$; $y^{-2} = \left(\dfrac{1}{4}\right)^2 = \dfrac{1}{16}$.

**Answer 14:**    A: None! Because $x = y$ or $x = -y$, all of the statements are true sometimes, but none is always true.

**Answer 15:**    C: $3^{-2} = \left(\dfrac{1}{3}\right)^2 = \dfrac{1}{9}$.

**Answer 16:**    B: $x^{\frac{3}{4}} = 8$; $x = \left(x^{\frac{3}{4}}\right)^{\frac{4}{3}} = 8^{\frac{4}{3}} = \left(\sqrt[3]{8}\right)^4 = 2^4 = 16$.

**Answer 17:**    C: We can ignore the minus sign, because both sides must be positive. $2^{6n} = 8^{n+4} = 2^{3(n+4)}$. So $6n = 3n + 12$, and $n = 4$.

This chapter seems the only good place to introduce the topic of **scientific notation**. On your calculator, if you see 3.176 and to the right of it you see 27, the number is written in scientific notation. It means $3.176 \times 10^{27}$.

A number in scientific notation is of the form $n \times 10^m$, where $n$ is at least 1 but less than 10.

**Example 2:** Write 567,000 in scientific notation.

**Solution:** Put the decimal place after the first digit 5.67; then count the number of places to the right (of 5) to the end of the original number. This is the power of 10, or $m$, so the answer is $5.67 \times 10^5$. It is a positive exponent because we are counting to the right and because the original number is larger than 10.

**Example 3:** Write 0.000000097 in scientific notation.

**Solution:** Again, put the decimal after the first digit, 9.7; then count, to the left, the number of places from the decimal point after the 9 to where it was in the original number. It is 8 places; the exponent of 10 is negative here because it is to the left and the entire number is less than 1. The answer is $0.000000097 = 9.7 \times 10^{-8}$.

**Example 4:** Compute the following in scientific notation: $\dfrac{60,000 \times .008}{.00002 \times 200,000,000}$

**Solution:** Write all the numbers in scientific notation : $\dfrac{6 \times 10^4 \times 8 \times 10^{-3}}{2 \times 10^{-5} \times 2 \times 10^8}$.

Separate the number parts from the powers of ten. As before, bring negative exponents in the top to the bottom and the bottom to the top, making all of them positive. Then do the math.

$$\frac{6 \times 8}{2 \times 2} \times \frac{10^4 \times 10^5}{10^3 \times 10^8} = 12 \times \frac{10^9}{10^{11}} = 12 \times 10^{-2} = 1.2 \times 10^1 \times 10^{-2} = 1.2 \times 10^{-1}$$

# Chapter 3 Quiz

Simplify the expressions in questions 1–5.

1. $(-3x^{20}y^{30})(-10x^4y)$

2. $\dfrac{(4x^3)^2(10x^4)^3}{(2x^3)^5}$

3. $x^{3a+2}x^{4a+7}$

4. $6^86^6$

5. $(4x)^0 + 4x^0 + 4^0x^0$

Evaluate the expressions in questions 6–10.

6. $\dfrac{4^{-3}}{2^{-5}}$

7. $8^{-\frac{5}{3}}$

8. $\dfrac{1}{100^{-\frac{7}{2}}}$

9. $3^{-3} - \dfrac{1}{4^{-\frac{3}{2}}}$

10. $4^64^{-8}4^{\frac{5}{2}}$

Simplify the expressions in questions 11–14.

11. $\left(\dfrac{(m^4)}{(m^{-3})}\right)^{-5}$

12. $\dfrac{4x^8y^{-4}z^{-5}}{12x^{-2}y^{-1}z^5}$

13. $(4xy^{-3}z^{-5})(-5x^{-5}y^2z^7)$

14. $x^{\frac{3}{2}}x^{\frac{2}{3}}x^{-\frac{1}{6}}$

15. Simplify $3^n + 3^n + 3^n$.

16. If $x^{-2} = \dfrac{1}{16}$, $x^3 = ?$

17. If $4^{3x+2} = 8^{4x+3}$, $x = ?$

18. Simplify $\dfrac{(x^m)^{m+1}}{x^m}$.

19. Simplify $(4xy^{-4}z^5a^{-6}b^{-1})^{-5}$.

20. In scientific notation, $\dfrac{(.0008)(800,000,000)}{.002} =$

21. Write the expression $4xy - 7y - y^2$ in terms of $x$ if $y = -2x$, and simplify.

22. Write the expression $4y^3 z^{10}$ in terms of $x$ if $y = 10x^2$ and $z = x^5$, and simplify.

## Answers to Chapter 3 Quiz

1. $30x^{24}y^{31}$.

2. $\dfrac{(16x^6)(1000x^{12})}{32x^{15}} = 500x^3$.

3. $x^{7a+9}$.

4. $6^{15}$.

5. $1 + 4 + 1 = 6$.

6. $\dfrac{2^5}{4^3} = \dfrac{32}{64} = \dfrac{1}{2}$.

7. $8^{-\frac{5}{3}} = \dfrac{1}{8^{\frac{5}{3}}} = \dfrac{1}{\left(\sqrt[3]{8}\right)^5} = \dfrac{1}{2^5} = \dfrac{1}{32}$.

8. $\dfrac{1}{100^{-\frac{7}{2}}} = 100^{\frac{7}{2}} = \left(\sqrt{100}\right)^7 = 10,000,000$.

9. $3^{-3} - \dfrac{1}{4^{-\frac{3}{2}}} = \dfrac{1}{27} - 8 = -7\dfrac{26}{27}$.

10. $4^{\frac{1}{2}} = 2$.

11. $\dfrac{m^{-20}}{m^{15}} = \dfrac{1}{m^{35}}$.

12. $\dfrac{4x^8x^2y^1}{12y^4z^5z^5} = \dfrac{x^{10}}{3y^3z^{10}}$.

13. $-\dfrac{20z^2}{x^4y}$.

14. $x^2$ (just add the exponents).

15. $1 \times 3^n + 1 \times 3^n + 1 \times 3^n = 3 \times 3^n = 3^1 3^n = 3^{n+1}$.

16. $\dfrac{1}{x^2} = \dfrac{1}{16}$; $x^2 = 16$; so $x = \pm 4$; and $x^3 = \pm 64$.

17. $4^{3x+2} = 8^{4x+3}$; $(2^2)^{3x+2} = (2^3)^{4x+3}$; so $2(3x+2) = 3(4x+3)$; $x = -\dfrac{5}{6}$.

18. $\dfrac{x^{m^2+m}}{x^m} = \dfrac{x^{m^2}x^m}{x^m} = x^{m^2}$.

19. $4^{-5}x^{-5}y^{20}z^{-25}a^{30}b^5 = \dfrac{a^{30}b^5y^{20}}{1{,}024x^5z^{25}}$. (It is prettier to write letters alphabetically and so that you don't forget one, but any order is correct!)

20. $\dfrac{8 \times 10^{-4} \times 8 \times 10^8}{2 \times 10^{-3}} = \dfrac{8 \times 8}{2} \times 10^{-4}10^8 10^3 = 32 \times 10^7 = 3.2 \times 10^8$.

21. $4x(-2x) - 7(-2x) - (-2x)^2 = -8x^2 + 14x - 4x^2 = -12x^2 + 14x$.

22. $4(10x^2)^3(x^5)^{10} = 4(1{,}000x^6)(x^{50}) = 4{,}000x^{56}$. Large numbers do not necessarily make a problem hard!

Now let's go to a radical chapter.

**Answer to "Bob Asks":** 31!!! You start the time from the first ring. So there are two rings in 1 second, 3 rings in 2 seconds, 6 rings in 5 seconds, and 31 rings in 30 seconds.

# CHAPTER 4: *The Most Radical Chapter of All*

*"If one hamburger and one hot dog cost $4.00, and one hamburger and three hot dogs cost $7.00, how much does one hamburger cost?"*

**The** square root symbol ($\sqrt{\ }$) is probably the one symbol most people actually like, even for people who don't like math. How else can you explain the square root symbol on a business calculator? I have yet to find a use for it. Here are some basic facts about square roots that you should know.

1.  You should know the following square roots:

    $\sqrt{0} = 0$    $\sqrt{1} = 1$    $\sqrt{4} = 2$    $\sqrt{9} = 3$    $\sqrt{16} = 4$    $\sqrt{25} = 5$

    $\sqrt{36} = 6$    $\sqrt{49} = 7$    $\sqrt{64} = 8$    $\sqrt{81} = 9$    $\sqrt{100} = 10$

    The numbers under the radicals (square root signs) are called perfect squares because their square roots are whole numbers.

2.  $\sqrt{2} \approx 1.4$ (actually it is 1.414 . . .), and $\sqrt{3} \approx 1.73$ (actually it is 1.732 . . . , the year George Washington was born).

3.  $\sqrt{\dfrac{a}{b}} = \dfrac{\sqrt{a}}{\sqrt{b}}$, so    $\sqrt{\dfrac{25}{9}} = \dfrac{5}{3}$    $\sqrt{\dfrac{7}{36}} = \dfrac{\sqrt{7}}{6}$    $\sqrt{\dfrac{45}{20}} = \sqrt{\dfrac{9}{4}} = \dfrac{3}{2}$.

4.  A method of simplification involves finding all the prime factors:

    $\sqrt{200} = \sqrt{(2)(2)(2)(5)(5)} = (2)(5)\sqrt{2} = 10\sqrt{2}$. We can also simplify by recognizing perfect squares: $\sqrt{200} = \sqrt{100 \times 2} = 10\sqrt{2}$. My choice is the first. You have to decide which method works best for you.

5.  Adding and subtracting radicals involves combining like radicals:

    $4\sqrt{7} + 5\sqrt{11} + 6\sqrt{7} - 9\sqrt{11} = 10\sqrt{7} - 4\sqrt{11}$

    You cannot combine unlike radicals.

6.  Multiplication of radicals follows this rule: $a\sqrt{b} \times c\sqrt{d} = ac\sqrt{bd}$.

    Therefore, $3\sqrt{13} \times 10\sqrt{7} = 30\sqrt{91}$, and

    $10\sqrt{8} \times 3\sqrt{10} = 10 \times 3\sqrt{(2 \times 2 \times 2) \times (2 \times 5)} = 3\sqrt{(2 \times 2) \times (2 \times 2) \times 5} =$

    $10 \times 3 \times 2 \times 2 \times \sqrt{5} = 120\sqrt{5}$.

7. If a single radical appears in the denominator of a fraction, rationalize the denominator (change it to a nonradical) by multiplying both numerator and denominator by the radical:

$$\frac{20}{7\sqrt{5}} = \frac{20}{7\sqrt{5}} \times \frac{\sqrt{5}}{\sqrt{5}} = \frac{20\sqrt{5}}{35} = \frac{4\sqrt{5}}{7} \text{ and } \frac{7}{\sqrt{45}} = \frac{7}{3\sqrt{5}} \times \frac{\sqrt{5}}{\sqrt{5}} = \frac{7\sqrt{5}}{15}$$

8. If the denominator has two terms, one or both of the terms having a square root, multiply the numerator and denominator by its **conjugate**, the same term with a different sign between.

So $\dfrac{3}{7 + \sqrt{5}} = \dfrac{3}{7 + \sqrt{5}} \times \dfrac{7 - \sqrt{5}}{7 - \sqrt{5}} = \dfrac{21 - 3\sqrt{5}}{49 - 5} = \dfrac{21 - 3\sqrt{5}}{44}$

This will make the denominator a whole number because $(a - b)(a + b) = a^2 - b^2$, and the difference of two squares is a whole number. More on this in Chapter 5.

9. If $c, d > 0$, $\sqrt{c} + \sqrt{d} > \sqrt{c + d}$. Why? If we square the right side, we get $c + d$. If we square the left side, we get $c + d + a$ middle term $(2\sqrt{cd})$.

10. You should know the following cube roots:

$$\sqrt[3]{1} = 1 \qquad \sqrt[3]{8} = 2 \qquad \sqrt[3]{27} = 3 \qquad \sqrt[3]{64} = 4 \qquad \sqrt[3]{125} = 5$$

$$\sqrt[3]{-1} = -1 \qquad \sqrt[3]{-8} = -2 \qquad \sqrt[3]{-27} = -3 \qquad \sqrt[3]{-64} = -4 \qquad \sqrt[3]{-125} = -5$$

Cube roots can be simplified in the same way as square roots. For example, $\sqrt[3]{32} = \sqrt[3]{(2 \times 2 \times 2) \times 2 \times 2} = 2\sqrt[3]{4}$.

**Note**   *When I went to school, everyone knew that the cube root of 1,728 is 12. We knew that 12 inches is equivalent to 1 foot and $12^3 = 1,728$ inches is equivalent to 1 cubic foot.*

**Note**   *If you had $4^{th}$ roots, you would try to find a number that appears as a product four times. For example, $\sqrt[4]{48} = \sqrt[4]{(2 \times 2 \times 2 \times 2) \times 3} = 2\sqrt[4]{3}$.*

11. Adding and subtracting cube roots follows item 5. For example, $3\sqrt[3]{9} + 7\sqrt[3]{9} - \sqrt[3]{2} = 10\sqrt[3]{9} - \sqrt[3]{2}$. You cannot combine unlike radicals.

12. Multiplication and division of cube roots follows item 6. Here are some examples.

$$10\sqrt[3]{6} \times 5\sqrt[3]{45} = 50\sqrt[3]{2 \times (3 \times 3 \times 3) \times 5} = 150\sqrt[3]{10}.$$

$$\sqrt[3]{125} \div \sqrt[3]{1728} = \frac{\sqrt[3]{5 \times 5 \times 5}}{\sqrt[3]{12 \times 12 \times 12}} = \frac{5}{12}.$$

$$\sqrt[3]{67} \div \sqrt[3]{64} = \frac{\sqrt[3]{67}}{\sqrt[3]{4 \times 4 \times 4}} = \frac{\sqrt[3]{67}}{4}.$$

13. The square root varies according to the value of the radicand:

If $a > 1$, $a > \sqrt{a}$. For example, $9 > \sqrt{9}$.

If $a = 1$, $a = \sqrt{a}$ because the square root of 1 is 1.

If $0 < a < 1$, $a < \sqrt{a}$. When we take the square root of a positive number, it becomes closer to 1 than the number, so $\sqrt{\dfrac{1}{4}} = \dfrac{1}{2} > \dfrac{1}{4}$.

If $a = 0$, $a = \sqrt{a}$ because the square root of 0 is 0.

If $a > 0$ and $\sqrt{a} < 1$, then $\dfrac{1}{\sqrt{a}} > 1$. Also, if $\sqrt{a} > 1$, then $\dfrac{1}{a} < 1$.

14. The square root or any even root of a negative number is imaginary (undefined).

15. The cube root (or any odd root) of a positive number is positive, of a negative number is negative, and any root of 0 is 0. All are always defined (not imaginary).

**Note**  $\sqrt{9} = 3$, $-\sqrt{9} = -3$, but $\sqrt{-9}$ is imaginary. The equation $x^2 = 9$ has two solutions, $\pm\sqrt{9}$, or $\pm 3$, which stands for both $+3$ and $-3$.

Let's do some practice problems.

**Example 1:**  Simplify the following radical expressions:

| Problem | Solution |
|---|---|
| a. $\sqrt{45}$ | $3\sqrt{5}$ |
| b. $5\sqrt{72}$ | $30\sqrt{2}$ |
| c. $4\sqrt{2} + 5\sqrt{3} - \sqrt{2} - 8\sqrt{3}$ | $3\sqrt{2} - 3\sqrt{3}$ or $3\left(\sqrt{2} - \sqrt{3}\right)$ |
| d. $5\sqrt{8} + 6\sqrt{27} + 9\sqrt{12}$ | $10\sqrt{2} + 36\sqrt{3}$ |
| e. $\left(3\sqrt{7}\right)\left(5\sqrt{11}\right)$ | $15\sqrt{77}$ |
| f. $\left(10\sqrt{21}\right)\left(4\sqrt{3}\right)$ | $120\sqrt{7}$ |
| g. $\left(4\sqrt{24}\right)\left(5\sqrt{8}\right)$ | $160\sqrt{3}$ |
| h. $\sqrt{\dfrac{81}{4}}$ | $\dfrac{9}{2}$ or $4\dfrac{1}{2}$ |
| i. $\sqrt{\dfrac{11}{49}}$ | $\dfrac{\sqrt{11}}{7}$ lucky answer |
| j. $\sqrt{\dfrac{27}{75}}$ | $\dfrac{3}{5}$ |
| k. $\sqrt{\dfrac{10}{72}}$ | $\dfrac{\sqrt{5}}{6}$ |

l.  $\dfrac{3}{\sqrt{5}}$                                    $\dfrac{3\sqrt{5}}{5}$

m. $\dfrac{32}{\sqrt{8}}$                                   $8\sqrt{2}$

n.  $\dfrac{4}{\sqrt{11} - 2}$                          $\dfrac{4\left(\sqrt{11} + 2\right)}{7}$

o.  $\dfrac{\sqrt{2}}{\sqrt{11} + \sqrt{5}}$                    $\dfrac{\sqrt{22} - \sqrt{10}}{6}$

**Q** **Let's try a few multiple-choice exercises.**

**Exercise 1:**  $\left(\sqrt{12} + \sqrt{27}\right)^{2} =$

A. 15                     D. 225

B. 39                     E. 675

C. 75

**Exercise 2:**  Suppose $0 < a < 1$.

I:       $a^2 > \sqrt{a}$

II:      $\sqrt{a} > \sqrt{a^3}$

III:     $\sqrt{a} > \dfrac{1}{\sqrt{a^7}}$

A. I is correct            D. I and III are correct

B. II is correct           E. II and III are correct

C. III is correct

**Exercise 3:**  $c = \left(\dfrac{1}{17}\right)^{2} - \sqrt{\dfrac{1}{17}}$. Which answer choice is true for $c$?

A. $c < -2$               D. $0 < c < 1$

B. $-2 < c < -1$          E. $1 < c < 2$

C. $-1 < c < 0$

**Exercise 4:** $0 < m < 1$. Arrange in order, smallest to largest, $a = \dfrac{1}{m}$, $b = \dfrac{1}{m^2}$, $c = \dfrac{1}{\sqrt{m}}$.

A. $a < b < c$      D. $b < a < c$

B. $a < c < b$      E. $c < a < b$

C. $b < c < a$

**Exercise 5:** If $\sqrt[3]{-87} = x$:

A. $-10 < x < -9$      D. $-4 < x < -3$

B. $-9 < x < -8$      E. $x$ is undefined

C. $-5 < x < -4$

**Exercise 6:** Simplified, $\dfrac{\sqrt{m}}{\sqrt{m} - \sqrt{n}} =$

A. $\dfrac{m - \sqrt{n}}{m - n}$      D. $\dfrac{m - \sqrt{mn}}{m + n}$

B. $\dfrac{m + \sqrt{n}}{m + n}$      E. $1$

C. $\dfrac{m + \sqrt{mn}}{m - n}$

**A** **Let's look at the answers.**

**Answer 1:** C: $\sqrt{12} = \sqrt{2 \times 2 \times 3} = 2\sqrt{3}$ and $\sqrt{27} = \sqrt{3 \times 3 \times 3} = 3\sqrt{3}$. Adding, we get $5\sqrt{3}$. Squaring, we get $25\sqrt{9} = 25 \times 3 = 75$.

**Answer 2:** B: Let's look at the statements one by one.

**Statement I:** If we square a number between 0 and 1, we make it closer to 0. If we take the square root of the same number, we make it closer to 1. Statement I is wrong.

**Statement II:** From the previous chapter, if $0 < a < 1$, $a > a^3$. So are its square roots. So statement II is true.

**Statement III:** $\sqrt{a} < 1$. $a^7 < 1$. So $\sqrt{a^7} < 1$. Then $\dfrac{1}{\sqrt{a^7}} > 1$. Statement III is false.

**Answer 3:**    C: $\frac{1}{17}$ squared is less than $\frac{1}{17}$, the square root is more than $\frac{1}{17}$, and both numbers are between 0 and 1. When we subtract a larger from a smaller number, the sign is negative, and here the numbers are fractions, so $-1 < c < 0$.

**Answer 4:**    E: Often, these comparison problems are figured out by using an example. If we take $m = \frac{1}{4}$, we see that $\sqrt{m} > m > m^2$. That makes $\frac{1}{\sqrt{m}} < \frac{1}{m} < \frac{1}{m^2}$, or $c < a < b$.

**Answer 5:**    C: Because $\sqrt[3]{-125} = -5$ and $\sqrt[3]{-64} = -4$, and because $-87$ is between $-125$ and $-64$, the cube root of $-87$ must be between $-5$ and $-4$.

**Answer 6:**    C: $\frac{\sqrt{m}}{\sqrt{m} - \sqrt{n}} \times \frac{\sqrt{m} + \sqrt{n}}{\sqrt{m} + \sqrt{n}} = \frac{m + \sqrt{mn}}{m - n}$. Note that answer choice E couldn't be correct because if the fraction was not equal to 1 before simplification, it couldn't be equal to 1 after simplification.

In many colleges, what you need to know about radicals and how much you need to know in taking further math courses is what was just shown. However, some colleges require more. Here we present some examples of additional facts about radicals from the beginning of this chapter. They involve letters and cube roots, fourth roots, and so on. Following these examples will be some practice exercises involving radicals.

**Example 2:**    Simplify: $4\sqrt{27x^9}$.

**Solution:**    Write the expression using prime factors and group them in pairs where possible. $4\sqrt{(3 \times 3) \times 3 \times (x \times x) \times (x \times x) \times (x \times x) \times (x \times x) \times x} = 12x^4\sqrt{3x}$.

**Note**    *As mentioned in fact #4 at the beginning of this chapter, I think this is the easiest way to simplify radicals with terms in exponential form inside the radical.*

*Looking at this example, we see that $\sqrt{x^9} = x^4\sqrt{x}$. Think that the square root is the "2th" root. If you divide 2 into 9, you get 4 with remainder of 1, represented by the $x^4$ term outside the radical and the remaining x inside the radical.*

Use fact #4 to simplify the expressions in Examples 3–5.

**Example 3:**    Simplify: $4\sqrt{3x} + 5\sqrt{6y} + 8\sqrt{3x} - 7\sqrt{6y}$.

**Solution:**    $12\sqrt{3x} - 2\sqrt{6y}$.

**Example 4:**    Simplify: $5x\sqrt{3y} - 9x\sqrt{3y}$.

**Solution:**    $-4x\sqrt{3y}$

**Example 5:**    Simplify: $5\sqrt{8x^5} + 7x\sqrt{18x^3} + 10x^2\sqrt{32x}$.

**Solution:**    The terms can be rewritten as $10x^2\sqrt{2x} + 21x^2\sqrt{2x} + 40x^2\sqrt{2x} = 71x^2\sqrt{2x}$.

**Example 6:**    Multiply and simplify: $5a^2b^3\sqrt{6a^4b^5} \times 2a^2\sqrt{8a^7b}$.

**Solution:**    Use fact #6 for the multiplication: $10a^4b^3\sqrt{48a^{11}b^6} =$ $(10a^4b^3)(4a^5b^3\sqrt{3a}) = 40a^9b^6\sqrt{3a}$.

**Example 7:**    Simplify: $\dfrac{4a^3b^2}{\sqrt{12a^5b^9}}$.

**Solution:**    Use fact #7 to eliminate the radical in the denominator (rationalize the fraction): $\dfrac{4a^3b^2}{2a^2b^4\sqrt{3ab}} \times \dfrac{\sqrt{3ab}}{\sqrt{3ab}} = \dfrac{4a^3b^2\sqrt{3ab}}{2a^2b^4(3ab)} = \dfrac{2\sqrt{3ab}}{3b^3}$.

**Note**    *In one of the more puzzling aspects of math, when you see the word "rationalize," you always rationalize the denominator. However, in calculus you will never rationalize the denominator. You will always rationalize the numerator. The reason is that before calculators, it was much more difficult, took longer, and was less accurate to determine the value of $\dfrac{1}{\sqrt{2}}$ than it was to determine the value of $\dfrac{\sqrt{2}}{2}$.*

*Whenever mathematicians make very slow changes in methodology, it is always good. Unfortunately, after 1985, the changes were the most rapid. (Translation: There is a lot of bad stuff around now.)*

**Example 8:**    Simplify: $\dfrac{\sqrt{a} + \sqrt{b}}{\sqrt{a} - \sqrt{b}}$.

**Solution:**    Use fact #8. $\dfrac{\sqrt{a} + \sqrt{b}}{\sqrt{a} - \sqrt{b}} \times \dfrac{\sqrt{a} + \sqrt{b}}{\sqrt{a} + \sqrt{b}} = \dfrac{a + 2\sqrt{ab} + b}{a - b}$.

**Example 9:**    Simplify: $\sqrt[3]{a^{20}}$.

**Solution:**    $a^6\sqrt[3]{a^2}$.

**Note**   *If we divide 3 into 20, we get 6 with a remainder of 2, which are the exponents of a.*

**Example 10:**  Simplify: $\sqrt[10]{x^{86}}$.

**Solution:**   $x^8\sqrt[10]{x^6}$

**Example 11:**  Simplify: $\sqrt[3]{32x^5}$.

**Solution:**   $x\sqrt[3]{(2\times2\times2\times)\times2\times2\times x^2} = 2x\sqrt[3]{4x^2}$.

**Example 12:**  Perform the following operations:

| Problem | Solution |
|---|---|
| **a.** Simplify: $\sqrt{16w^3x}$. | $4w\sqrt{wx}$ |
| **b.** Simplify: $\sqrt{24u^6v^9}$. | $2u^3v^4\sqrt{6v}$ |
| **c.** Simplify: $5x^2\sqrt{20x^{101}}$. | $10x^{52}\sqrt{5x}$ |
| **d.** Simplify: $\sqrt[3]{72z^{32}}$. | $2z^{10}\sqrt[3]{9z^2}$ |
| **e.** Simplify and combine: $5x^3\sqrt{28x^5} + 7x^4\sqrt{63x^3}$. | $10x^5\sqrt{7x} + 21x^5\sqrt{7x} = 31x^5\sqrt{7x}$ |
| **f.** Simplify: $\sqrt[3]{x^{100}}$. | $x^{33}\sqrt[3]{x}$ |
| **g.** Simplify: $\sqrt[5]{y^{77}}$. | $y^{15}\sqrt[5]{y^2}$ |
| **h.** Simplify: $10y\sqrt[3]{108y^{14}}$. | $30y^5\sqrt[3]{4y^2}$ |
| **i.** Multiply and simplify: $\left(2s^4t^{10}\sqrt{6s^3t}\right)\left(10s^2t^2\sqrt{2st^2}\right)$. | $40s^8t^{13}\sqrt{3t}$ |
| **j.** Rationalize: $\dfrac{4x^7}{\sqrt{72x^{21}}}$. | $\dfrac{\sqrt{2x}}{3x^4}$ |
| **k.** Rationalize: $\dfrac{4u}{7-\sqrt{u}}$. | $\dfrac{4u\left(7+\sqrt{u}\right)}{49-u}$ |
| **l.** Rationalize: $\dfrac{\sqrt{y}-\sqrt{z}}{\sqrt{y}+\sqrt{z}}$. | $\dfrac{y-2\sqrt{yz}+z}{y-z}$ |

Let's look at imaginary numbers.

# IMAGINARY (COMPLEX) NUMBERS

For this particular book, this is the right time to discuss imaginary (or complex) numbers. Even though this is only a few pages and for most people very simple, it is enough about the subject to take you at least past the entire basic calculus sequence in college, if not further.

Once upon a time, we had the counting numbers (natural numbers) 1, 2, 3, 4, . . . . But mathematicians wanted an answer to the equation $x + 6 = 6$. So 0 was invented. (Actually, 0 was invented in India around the seventh or eighth century. At that time, the Indians were the great mathematicians of the world. Today, India is still pretty darn good!) Then mathematicians wanted even more. They wanted the answer to the equation $x + 5 = 3$. So negative integers were invented. This was still not enough. Mathematicians wanted an answer to the equations $3x = 7$ and $5x = -2$. So rational numbers were invented. Still, they wanted an answer to the equation $x^2 = 41$. So irrational numbers, such as $\pm \sqrt{41}$, were invented. Finally, mathematicians wanted an answer to the equation $x^2 = -1$. Alas! There is no real number such that $x^2 = -1$. So they invented $i$ to stand for the "imaginary" number $\sqrt{-1}$. This was almost complete, but not quite enough. **Complex numbers** were invented. A **complex number** is of the form $a + bi$, where $a$ and $b$ are real numbers, and $i = \sqrt{-1}$.

Two complex numbers are equal if their real parts are equal and their imaginary parts are equal.

$$a + bi = c + di \text{ if } a = c \text{ and } b = d$$

Complex numbers cannot be graphed on a real graph. So instead of $x$ and $y$ axes, there is an $x$ and a $yi$ axis. On the axes below, we have graphed the point $-8 + 6i$.

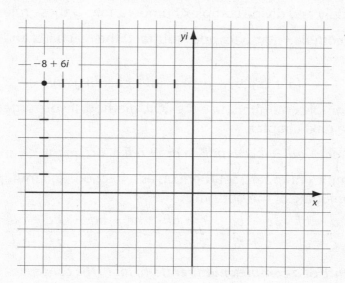

There are only a few things we need to know here about complex numbers.

The distance of the number $a + bi$ from the origin $(0 + 0i)$ is $|a + bi| = \sqrt{a^2 + b^2}$. This is also called its **magnitude**. It is similar to what we get for real numbers by using the distance formula.

**Example 13:**  What is the distance of the point $(6 - 8i)$ from the origin?

**Solution:**    $|6 - 8i| = \sqrt{6^2 + (-8)^2} = 10.$

> **Note**   *The distance is always a positive real number and i is not involved in the square root.*

We also need to know how to raise $i$ to any power. We see that $i = \sqrt{-1}$, so $i^2 = -1$. $i^3 = i^2 i = -i$; $i^4 = i^2 i^2 = (-1)(-1) = 1$; $i^5 = i^4 i = i$, etc. So $i$ is cyclic, and the cycle is 4. This means the value of $i$ raised to a power repeats itself every fourth power.

**Example 14:**  Simplify $i^{1359}$.

**Solution:**    To find what part of the $i^n$ cycle this is, we divide 1,359 by 4.

We are not interested in the answer—only the remainder.

The answer is 339 R 3. Because the remainder is 3, the answer is the same as $i^3$, or $-i$.

To summarize, to find the value of $i$ raised to any power, we divide the exponent by 4. If the remainder is 1, the answer is $i$. If the remainder is 2, the answer is $-1$. If the remainder is 3, the answer is $-i$. If the remainder is 0, the answer is 1.

To add, subtract, multiply and divide complex numbers, follow these rules:

1.   To add (subtract) complex numbers, add (subtract) the real parts and add (subtract) the imaginary parts.

$$(a + bi) \pm (c + di) = (a + c) \pm (b + d)i.$$

2.   To multiply two complex numbers, use the FOIL method. If you forget what the FOIL method is, see the next chapter.

$$(a + bi)(c + di) = ac + adi + bci + bdi^2 = ac + adi + bci + bd(-1) = ac - bd + (ad + bc)i$$

3.   To divide two complex numbers, set them up as a fraction, and multiply the top and bottom by the conjugate of the bottom.

$$\frac{a + bi}{c + di} = \frac{a + bi}{c + di} \times \frac{c - di}{c - di} = \frac{ac + bd + i(bc - ad)}{c^2 + d^2}$$

> **Note**   *The **conjugate of a complex number** is similar to the concept of conjugates with radicals, namely, the same expression with a different sign between the terms.*

> **Note**   *The final denominator is always a positive real number. It has a plus sign between the terms because it is $c^2 - (di)^2 = c^2 - (-d^2) = c^2 + d^2$.*

**Example 15:** Let $Y = 8 + 2i$ and $Z = 6 + 5i$. Find **a.** $Y + Z$; **b.** $Y - Z$; **c.** $YZ$; **d.** $\dfrac{Y}{Z}$.

**Solution:**

**a.** $Y + Z = 14 + 7i$

**b.** $Y - Z = 2 - 3i$

**c.** $YZ = 48 + 40i + 12i + 10i^2 = 38 + 52i$

**d.** $\dfrac{Y}{Z} = \dfrac{8 + 2i}{6 + 5i} \times \dfrac{6 - 5i}{6 - 5i} = \dfrac{48 - 40i + 12i - 10i^2}{36 - 30i + 30i - 25i^2} = \dfrac{58 - 28i}{61}$

$\quad\quad = \dfrac{58}{61} - \dfrac{28}{61}i$

It is essential when taking the square root of a negative number that you write the square root of a negative number in terms of *i* before proceeding.

Let us see why. When we wrote the rules for radicals, we were assuming positive or nonnegative numbers. In symbols,

$$\sqrt{a} \times \sqrt{b} = \sqrt{ab} \text{ if } a, b \geq 0, \text{ and } \sqrt{\dfrac{a}{b}} = \dfrac{\sqrt{a}}{\sqrt{b}} \text{ if } a \geq 0 \text{ and } b \geq 0.$$

Let us see what happens if this is ignored.

$$\sqrt{-2} \times \sqrt{-8} = i\sqrt{2} \times i\sqrt{8} = i^2\sqrt{16} = -4.$$

However, if we wrongly multiplied $-2$ by $-8$, we would get $+16 = +4$. Still not convinced? I will "wrongly" prove that $-1 = +1$.

| <u>Step</u> | <u>Reason</u> |
|---|---|
| 1. $-1 = -1$ | Any number equals itself (known as the reflexive property). |
| 2. $\dfrac{1}{-1} = \dfrac{-1}{1}$ | A positive divided by a negative is the same as a negative divided by a positive. |
| 3. $\sqrt{\dfrac{1}{-1}} = \sqrt{\dfrac{-1}{1}}$ | Square roots of equals are equal. |
| 4. $\dfrac{\sqrt{1}}{\sqrt{-1}} = \dfrac{\sqrt{-1}}{\sqrt{1}}$ | Rule #4: $\sqrt{\dfrac{x}{y}} = \dfrac{\sqrt{x}}{\sqrt{y}}$. |
| 5. $\dfrac{1}{i} = \dfrac{i}{1}$ | $\sqrt{1} = 1$ and $\sqrt{-1} = i$. |
| 6. $i \times \dfrac{1}{i} = i \times \dfrac{i}{1}$ | If you multiply equals by equals, their products are equal. |
| 7. $1 = -1$ | $i^2 = -1$. |

Did I just prove the impossible? Does 1 really equal $-1$? If you think it's true, you can give me one million dollars, but you still owe me one million since $1 = -1$.

Of course it's not true. See if you can figure out why. (See the end of the chapter if you can't.)

**Example 16:** Rewrite in terms of $i$:

| Problem | Solution |
|---|---|
| **a.** $\sqrt{-9}$ | $3i$ |
| **b.** $-7 + \sqrt{-16}$ | $-7 + 4i$ |
| **c.** $10 + \sqrt{-32}$ | $10 + 4i\sqrt{2}$ |

This will blow your mind. If you have the set of complex numbers $a + bi$, where $a$ and $b$ are real integers, the number 2 is no longer a prime! The reason is because $2 = (1 + i)(1 - i)$. The number 3 is still a prime, but 5 and 17 are not primes! See the end of the chapter to see why this is true. The more math you know, the more interesting it gets. I love math!

## Chapter 4 Quiz

1. Simplify: $11\sqrt{84}$.
2. Simplify: $4x^2 \sqrt{54x^7}$.
3. Simplify: $5\sqrt[3]{128}$.
4. Simplify: $4x\sqrt[4]{64x^{27}}$.
5. Simplify and combine: $4\sqrt{12} + 6\sqrt{18} + 8\sqrt{75}$.
6. Simplify and combine: $8z^4 \sqrt{8z^5} - 10z^3\sqrt{32z^7}$.
7. $\left(4\sqrt{11}\right)\left(7\sqrt{7}\right) =$
8. $4x\sqrt{8x} \times 5x^2\sqrt{3x^5} =$

Rationalize the denominator in questions 9 and 10:

9. $\dfrac{9a^9}{\sqrt{27a^{13}}} =$

10. $\dfrac{\sqrt{7}}{\sqrt{7} + \sqrt{3}} =$

11. If $5 + 6i = a + 3 + (b - 7)i$, $a$ and $b =$

12. $i^{232} =$

13. $3(4i - 7) + 5i(3i - 2) =$

14. $i + i^2 + i^3 + i^4 =$

15. The magnitude of $-6 - 7i =$

## Answers to Chapter 4 Quiz

1. $11 \times 2\sqrt{21} = 22\sqrt{21}$.

2. $(4x^2)(3x^3)\sqrt{6x} = 12x^5\sqrt{6x}$.

3. $5\sqrt[3]{64 \times 2} = 20\sqrt[3]{2}$.

4. $(4x)(2x^6)\sqrt[4]{4x^3} = 8x^7\sqrt[4]{4x^3}$.

5. $4 \times 2\sqrt{3} + 6 \times 3\sqrt{2} + 8 \times 5\sqrt{3} = 48\sqrt{3} + 18\sqrt{2}$.

6. $8z^4 \times 2z^2\sqrt{2z} - 10z^2 \times 4z^3\sqrt{2z} = -24z^6\sqrt{2z}$.

7. $28\sqrt{77}$.

8. $4x \times 2\sqrt{2x} \times 5x^2 \times x^2\sqrt{3x} = 40x^6\sqrt{6}$.

9. $\dfrac{9a^9 \times \sqrt{3a}}{3a^6\sqrt{3a} \times \sqrt{3a}} = a^2\sqrt{3a}$.

10. $\dfrac{\left(\sqrt{7}\right)\left(\sqrt{7} - \sqrt{3}\right)}{\left(\sqrt{7} + \sqrt{3}\right)\left(\sqrt{7} - \sqrt{3}\right)} = \dfrac{7 - \sqrt{21}}{4}$.

11. $a + 3 = 5$ and $b - 7 = 6$; so $a = 2$ and $b = 13$.

12. $\dfrac{232}{4}$ has a remainder of 0; so the answer is the same as $i^0 = 1$.

13. $12i - 21 + 15i^2 - 10i = -21 - 15 + 2i = -36 + 2i$.

14. $i - 1 - i + 1 = 0$.

15. $|-6 - 7i| = \sqrt{(-6)^2 + (-7)^2} = \sqrt{85}$.

The answer to the wrong proof ($1 = -1$) is that you can't, _can't_, _can't_ do step 4 since you have two square roots of negative numbers! Division of radicals works only for nonnegative numbers. It is illegal (a sick bird).

$5 = (2 - i)(2 + i)$ and $17 = (4 + i)(4 - i)$.

Now let's do some basic algebra that was taught in Algebra 1 when I went to school several hundred years ago.

**Answer to "Bob Asks":** $B + D = 4$ and $B + 3D = 7$. Subtracting the second equation from the first, we get $2D = 3$. So a hot dog $D = \$1.50$. From the first equation, a burger $B = \$2.50$.

"*I like English puzzles also. Can you find a word with the letters WKW in it in that order?*"

**Algebraic** manipulative skills such as those in this chapter are areas that high schools have tended to de-emphasize since 1985. This chapter is extremely important, both for the COMPASS and success in math college precalculus and calculus courses that you may need to take. If I needed a quick placement test for higher math courses, I might just give a test from this chapter only. As it turns out, the COMPASS does have questions from this chapter. This is the chapter that you may need to study the most. You must know this chapter really well if you are going forward in math-related courses and want to succeed.

## COMBINING LIKE TERMS

**Like terms** are terms with the same letter combination (or no letter). The same letter must also have the same exponents.

**Example 1:** Are the following terms like or unlike?

    **a.** $4x$ and $-5x$

    **b.** $4x$ and $4x^2$

    **c.** $xy^2$ and $x^2y$

**Solutions:** **a.** $4x$ and $-5x$ are like terms even though their numerical coefficients are different. We are concerned only with the same exponents.

    **b.** $4x$ and $4x^2$ are unlike terms. They have different exponents.

    **c.** $xy^2$ and $x^2y$ are unlike; $xy^2 = xyy$ and $x^2y = xxy$. Again, different exponents.

**Combining like terms** means adding or subtracting their numerical coefficients; exponents are unchanged. Unlike terms cannot be combined.

**Example 2:**    Simplify:

| Problem | Solution |
|---|---|
| **a.** $3m + 4m + m =$ | $8m$ |
| **b.** $8m + 2n + 7m - 7n =$ | $15m - 5n$ |
| **c.** $3x^2 + 4x - 5 - 7x^2 - 4x + 8 =$ | $-4x^2 + 3$ |

## DISTRIBUTIVE LAW

The **distributive law** states:

$$a(x + y) = ax + ay$$

**Example 3:**    Perform the indicated operations:

| Problem | Solution |
|---|---|
| **a.** $4(3x - 7) =$ | $12x - 28$ |
| **b.** $5(2a - 5b + 3c) =$ | $10a - 25b + 15c$ |
| **c.** $3x^4(7x^3 - 4x - 1) =$ | $21x^7 - 12x^5 - 3x^4$ |
| **d.** $4(3x - 7) - 5(4x - 2) =$ | $12x - 28 - 20x + 10 = -8x - 18$ |

## BINOMIAL PRODUCTS

A **binomial** is a two-term expression, such as $x + 2$. We use the **FOIL method** to multiply a binomial by a binomial. FOIL is an acronym for *First, Outer, Inner, Last*. This means to multiply the first two terms, then the outer terms, then the inner terms, and finally the last two terms.

**Example 4:**    Multiply $(x + 4)(x + 6)$

**Solution:**

Multiplying, we get $x^2 + 6x + 4x + 24 = x^2 + 10x + 24$.

You should know the following common binomial products:

$$(a + b)(a - b) = a^2 - b^2$$

$$(a - b)(a - b) = a^2 - 2ab + b^2$$

$$(a + b)(a + b) = a^2 + 2ab + b^2$$

**Note** *For a perfect square $(a + b)^2$, the first term of the resulting trinomial is the first term squared $(a^2)$, and the third term of the resulting trinomial is the last term squared $(b^2)$. The middle term is twice the product of the two terms of the binomial $(2ab)$, so*

$$(a + b)^2 = a^2 + 2ab + b^2$$

**Example 5:** Perform the indicated multiplications:

| Problem | Solution |
|---|---|
| **a.** $(x + 7)(x + 4) =$ | $x^2 + 4x + 7x + 28 = x^2 + 11x + 28$ |
| **b.** $(x - 5)(x - 2) =$ | $x^2 - 7x + 10$ |
| **c.** $(x + 6)(x - 3) =$ | $x^2 + 3x - 18$ |
| **d.** $(x + 6)(x - 8) =$ | $x^2 - 2x - 48$ |
| **e.** $(x + 5)(x - 5) =$ | $x^2 - 5x + 5x - 25 = x^2 - 25$ |
| **f.** $(x + 5)^2 =$ | $(x + 5)(x + 5) = x^2 + 10x + 25$ |
| **g.** $(x - 10)^2 =$ | $x^2 - 20x + 100$ |
| **h.** $(2x + 5)(3x - 10) =$ | $6x^2 - 5x - 50$ |
| **i.** $3(x + 4)(x + 5) =$ | $3(x^2 + 9x + 20) = 3x^2 + 27x + 60$ |
| **j.** $7(4x + 3)(4x - 3) =$ | $7(16x^2 - 9) = 112x^2 - 63$ |
| **k.** $(a + 4)(a + 7)$ | $a^2 + 11a + 28$ |
| **l.** $(b + 5)(b + 6)$ | $b^2 + 11b + 30$ |
| **m.** $(c + 1)(c + 9)$ | $c^2 + 10c + 9$ |

**n.** $(d + 4)(d + 8)$ $\qquad\qquad$ $d^2 + 12d + 32$

**o.** $(e + 11)(e + 10)$ $\qquad$ $e^2 + 21e + 110$

**p.** $(f - 6)(f - 2)$ $\qquad\quad$ $f^2 - 8f + 12$

**q.** $(g - 10)(g - 20)$ $\qquad$ $g^2 - 30g + 200$

**r.** $(h - 4)(h - 3)$ $\qquad\quad$ $h^2 - 7h + 12$

**s.** $(j - 1)(j - 7)$ $\qquad\quad$ $j^2 - 8j + 7$

**t.** $(k - 3)(k - 5)$ $\qquad\quad$ $k^2 - 8k + 15$

**u.** $4\sqrt{2}\left(5\sqrt{7} + 6 + \sqrt{8} + 5\sqrt{14}\right)$ $\quad$ $20\sqrt{14} + 24\sqrt{2} + 16 + 40\sqrt{7}$

**v.** $5x^2\sqrt{3x}\left(6x^4\sqrt{7x^6} + 4x\sqrt{3x}\right)$ $\quad$ $30x^9\sqrt{21x} + 60x^4$ ?

**w.** $\left(4\sqrt{2} + 6\sqrt{5}\right)\left(8\sqrt{2} + 10\sqrt{5}\right)$ $\quad$ $364 + 88\sqrt{10}$

**Example 6:** Perform the indicated multiplications:

<u>Problem</u> $\qquad\qquad\qquad$ <u>Solution</u>

**a.** $(k + 5)(k - 2)$ $\qquad\quad$ $k^2 + 3k - 10$

**b.** $(m + 5)(m - 8)$ $\qquad$ $m^2 - 3m - 40$

**c.** $(n - 6)(n + 2)$ $\qquad\quad$ $n^2 - 4n - 12$

**d.** $(p - 8)(p + 10)$ $\qquad$ $p^2 + 2p - 80$

**e.** $(q - 5r)(q + 2r)$ $\qquad$ $q^2 - 3qr - 10r^2$

**f.** $(s + 3)^2$ $\qquad\qquad\quad$ $s^2 + 6s + 9$

**g.** $(t - 4)^2$ $\qquad\qquad\quad$ $t^2 - 8t + 16$

**h.** $(3u + 5)^2$ $\qquad\qquad$ $9u^2 + 30u + 25$

**i.** $(5v - 4)^2$ $\qquad\qquad$ $25v^2 - 40v + 16$

**j.** $(ax + by)^2$ $\qquad$ $a^2x^2 + 2abxy + b^2y^2$

**k.** $(be - ma)^2$ $\qquad$ $b^2e^2 - 2\,beam + m^2a^2$

**l.** $(w + x)(w - x)$ $\qquad$ $w^2 - x^2$

**m.** $(a - 11)(a + 11)$ $\qquad$ $a^2 - 121$

**n.** $(am - 7)(am + 7)$ $\qquad$ $a^2m^2 - 49$

**o.** $(a^2b + c)(a^2b - c)$ $\qquad$ $a^4b^2 - c^2$

**p.** $3(x + 5)(x - 2)$ $\qquad$ $3x^2 + 9x - 30$

**q.** $-4(2x - 5)(3x - 4)$ $\qquad$ $-24x^2 + 92x - 80$

**r.** $x(2x - 5)(4x + 7)$ $\qquad$ $8x^3 - 6x^2 - 35x$

**s.** $5(x - 5)(x + 5)$ $\qquad$ $5x^2 - 125$

**t.** $\left(a\sqrt{b} + c\sqrt{d}\right)\left(e\sqrt{f} + g\sqrt{h}\right)$ $\quad$ $ae\sqrt{bf} + ag\sqrt{bh} + ce\sqrt{df} + cg\sqrt{dh}$

**u.** $4x\sqrt{7x}\left(4x\sqrt{7x} + 3x^3\sqrt{5x^{11}}\right)$ $\quad$ $112x^3 + 12x^{10}\sqrt{35}$

**v.** $\left(2\sqrt{3} + 4\sqrt{5}\right)^2$ $\qquad$ $92 + 16\sqrt{15}$

**w.** $\left(5\sqrt{u} + 3\sqrt{v}\right)^2$ $\qquad$ $25u + 30\sqrt{uv} + 9v$

**x.** Rationalize: $\dfrac{\sqrt{y} - \sqrt{z}}{\sqrt{y} + \sqrt{z}}$ $\qquad$ $\dfrac{y - 2\sqrt{yz} + z}{y - z}$

**Q Let's do a few multiple-choice exercises.**

**Exercise 1:** $x^2 - y^2 = 24;\ 3(x + y)(x - y) =$

A. 8 $\qquad$ D. 72

B. 24 $\qquad$ E. 13,824

C. 27

**Exercise 2:**    $x + y = m; x - y = \dfrac{1}{m}; x^2 - y^2 =$

A. $m^2$                    D. $\dfrac{1}{m}$

B. $m$                       E. $\dfrac{1}{m^2}$

C. $1$

**Exercise 3:**    $\left(x + \dfrac{1}{x}\right)^2 = 64; x^2 + \dfrac{1}{x^2} =$

A. $9$                       D. $65$

B. $62$                      E. $66$

C. $64$

**Exercise 4:**    $x^2 + y^2 = 20$ and $xy = -6$. Then $(x + y)^2 =$

A. $8$                       D. $26$

B. $14$                      E. $32$

C. $20$

**Exercise 5:**    $(x - y)^2 + 4xy =$

A. $x^2 + 8x + y^2$              D. $(x + y)^2$

B. $x^2 + 4x + y^2$              E. $x^2 + y^2$

C. $x^2 - y^2$

Ⓐ  **Let's look at the answers.**

**Answer 1:**    D: $3(x + y)(x - y) = 3(x^2 - y^2) = 3(24) = 72.$

**Answer 2:**    C: $(x + y)(x - y) = x^2 - y^2 = \dfrac{m}{1} \times \dfrac{1}{m} = 1.$

**Answer 3:**    B: $\left(x + \dfrac{1}{x}\right)\left(x + \dfrac{1}{x}\right) = x^2 + 2(x)\left(\dfrac{1}{x}\right) + \dfrac{1}{x^2} = x^2 + \dfrac{1}{x^2} + 2 = 64.$

So $x^2 + \dfrac{1}{x^2} = 64 - 2 = 62.$

**Answer 4:**    A: $(x + y)^2 = x^2 + 2xy + y^2 = x^2 + y^2 + 2xy = 20 + 2(-6) = 8$.

**Answer 5:**    D: $x^2 - 2xy + y^2 + 4xy = x^2 + 2xy + y^2 = (x + y)^2$.

Let's go on to factoring.

# FACTORING

**Factoring** is the reverse of the distributive law. There are three types of factoring you need to know: largest common factor, difference of two squares, and trinomial factorization.

If the distributive law says $x(y + z) = xy + xz$, then taking out the largest common factor says $xy + xy = x(y + z)$. Let's demonstrate a few factoring examples.

**Example 7:**    Factor the following problems:

| Problem | Solution | Explanation |
|---------|----------|-------------|
| a. $4x + 6y - 8$ | $2(2x + 3y - 4)$ | 2 is the largest common factor. |
| b. $8ax + 12ay - 40az$ | $4a(2x + 3y - 10z)$ | 4 is the largest common factor; $a$ is also a common factor. |
| c. $10a^4y^6z^3 - 15a^7y$ | $5a^4y(2y^5z^3 - 3a^3)$ | The largest common factor and the lowest power of each common variable is factored out: $a^4$ and $y$, but not $z$ because it is not in both terms. |
| d. $x^4y - xy^3 + xy$ | $xy(x^3 - y^2 + 1)$ | Factor out the lowest power of each common variable. Three terms in the original give three terms in parentheses. Note that $1 \times xy = xy$. |
| e. $9by + 12be + 4ye$ | prime | Some expressions cannot be factored. |

## Difference of Two Squares

Because $(a + b)(a - b) = a^2 - b^2$, factoring tells us that $a^2 - b^2 = (a + b)(a - b)$.

**Example 8:**

| Problem | Solution | Explanation |
|---|---|---|
| **a.** $x^2 - 25$ | $(x + 5)(x - 5)$ or $(x - 5)(x + 5)$ | Either order is OK. |
| **b.** $x^2 - 121$ | $(x + 11)(x - 11)$ | |
| **c.** $9a^2 - 25b^2$ | $(3a + 5b)(3a - 5b)$ | |
| **d.** $5a^3 - 20a$ | $5a(a^2 - 4) = 5a(a + 2)(a - 2)$ | Factor out the largest common factor first, then use the difference of two squares. |
| **e.** $x^4 - y^4$ | $(x^2 + y^2)(x^2 - y^2) =$ $(x^2 + y^2)(x + y)(x - y)$ | This is the difference of two squares where the square roots in the factors are also squares. Sum of two squares doesn't factor, but use the difference of two squares again. |

**Example 9:**    Factor completely:

| Problem | Solution |
|---|---|
| **a.** $12am + 18an$ | $6a(2m + 3n)$ |
| **b.** $6at - 18st + 4as$ | $2(3at - 9st + 2as)$ |
| **c.** $10ax + 15ae - 16ex$ | Prime |
| **d.** $18a^5c^6 - 27a^3c^8$ | $9a^3c^6(2a^2 - 3c^2)$ |
| **e.** $25a^4b^7c^9 - 75a^8b^9c^{10}$ | $25a^4b^7c^9(1 - 3a^4b^2c)$ |
| **f.** $a^4b^5 + a^7b - ab$ | $ab(a^3b^4 + a^6 - 1)$ |
| **g.** $9 - x^2$ | $(3 + x)(3 - x)$ |

      **h.** $x^4 - 36y^2$                   $(x^2 + 6y)\,(x^2 - 6y)$

      **i.** $2x^3 - 98x$                  $2x(x + 7)\,(x - 7)$

      **j.** $a^4 - 81b^2$                 $(a^2 + 9b)\,(a^2 - 9b)$

      **k.** $x^2 - 49$                    $(x + 7)\,(x - 7)$

      **l.** $5z^2 - 25$                   $5(z^2 - 5)$

      **m.** $a^4 - c^8$                  $(a^2 + c^4)\,(a + c^2)\,(a - c^2)$

      **n.** $2a^9 - 32a$                $2a(a^4 + 4)\,(a^2 + 2)\,(a^2 - 2)$

**Q**  **Let's do a couple more multiple-choice exercises.**

**Exercise 6:**   $8x + 6y = 30.$ Then $20x + 15y =$

      **A.** 15              **D.** 75

      **B.** 30              **E.** 150

      **C.** 60

**Exercise 7:**   $x^2 - 4 = 47 \times 43; x =$

      **A.** 41              **D.** 47

      **B.** 43              **E.** 49

      **C.** 45

**A**  **Let's look at the answers.**

**Answer 6:**   **D:** $8x + 6y = 2(4x + 3y) = 30;$ so $4x + 3y = 15.$ $20x + 15y = 5(4x + 3y)$
               $= 5 \times 15 = 75.$

**Answer 7:**   **C:** $x^2 - 4 = (x + 2)\,(x - 2) = 47 \times 43,$ or $(45 + 2)\,(45 - 2).$ So $x = 45.$

Later in this chapter as well as in later chapters, we will see more comparison questions. For now, let's do trinomial factoring.

## Factoring Trinomials

Factoring trinomials is a puzzle, a game, which is rarely done well in high school and even more rarely practiced. Let's learn the factoring game.

First, let's rewrite the first four parts of Example 5 backward and look at them.

**a.** $x^2 + 11x + 28 = (x + 7)(x + 4)$

**b.** $x^2 - 7x + 10 = (x - 5)(x - 2)$

**c.** $x^2 + 3x - 18 = (x + 6)(x - 3)$

**d.** $x^2 - 2x - 48 = (x + 6)(x - 8)$

Each term starts with $x^2$ ($= +1x^2$), so the first sign is $+$. We'll call the sign in front of the $x$ term the middle sign, and we'll call the sign in front of the number term the last sign.

Let's look at **a** and **b** above to state some rules of the game:

1.  If the last sign (in the trinomial) is $+$, then both signs (in the parentheses) must be the same. The reason? $(+) \times (+) = +$, and $(-) \times (-) = +$.

2.  Only if the last sign is $+$, look at the sign of the middle term. If it is $+$, both factors have a $+$ sign (as in **a**); if it is $-$, both factors have a $-$ sign (as in **b**).

3.  If the last sign is $-$, the signs in the two factors must be different (see **c** and **d**).

Now, let's play the game.

**Example 10:** Factor: $x^2 - 16x + 15$.

**Solution:**

1.  The last sign $+$ means both signs are the same. The middle sign $-$ means both are $-$.

2.  The only factors of $x^2$ are $(x)(x)$. Look at the number term 15. The factors of 15 are (3)(5) and (1)(15). So $(x - 5)(x - 3)$ and $(x - 15)(x - 1)$ are the only possibilities. We have chosen the first and last terms to be correct, so we need to check only the middle term. The first, $-8x$, is wrong; the second, $-16x$, is correct.

3.  The answer is $(x - 15)(x - 1)$. If neither worked, the trinomial couldn't be factored.

**Example 11:** Factor completely: $x^2 - 4x - 21$.

**Solution:**

1.  The last sign negative means the signs of the factors are different.

2.  $x^2 = x(x)$ and the factors of 21 are 7 and 3 or 21 and 1. We want the pair that totals the middle number $(-4)$, so $-7$ and 3 are correct.

Note that the larger factor (7) gets the minus sign because the middle number is negative.

3. The answer is $(x - 7)(x + 3)$.

 **Note**    *If you multiply the inner and outer terms and get the right number but the wrong sign, both signs in the parentheses must be changed.*

The game gets more complicated if the coefficient of $x^2$ is not 1.

**Example 12:**  Factor completely: $4x^2 + 4x - 15$.

**Solution:**

1. The last sign is $-$, so the signs of the factors must be different.

2. The factors of $4x^2$ are $(4x)(x)$ or $(2x)(2x)$.

3. The factors of 15 are 3 and 5, or 1 and 15. Let's write out all the possibilities and the resulting middle terms. We are looking for terms whose difference is $4x$.

   $(4x \quad 3)(x \quad 5)$; middle terms are $3x$ and $20x$, no way to get $4x$.

   $(4x \quad 5)(x \quad 3)$; middle terms are $5x$ and $12x$, and again, there is no way to get $4x$.

   $(4x \quad 15)(x \quad 1)$; middle terms are $15x$ and $4x$, wrong!

   $(4x \quad 1)(x \quad 15)$; middle terms are $1x$ and $60x$, all the way in the next county.

   $(2x \quad 1)(2x \quad 15)$; middle terms are $2x$ and $30x$, wrong again!

   $(2x \quad 3)(2x \quad 5)$; middle terms are $6x$ and $10x$, correct, whew!

4. The minus sign goes in front of the 3 and the plus sign goes in front of the 5 to get the middle coefficient of $+4$, so the answer is $(2x - 3)(2x + 5)$.

**Example 13:**  Factor completely: $3x^2 + 15x + 12$.

**Solution:**    Take out the common factor first.

$3x^2 + 15x + 12 = 3(x^2 + 5x + 4) = 3(x + 4)(x + 1)$.

**Example 14.**  Factor completely; coefficients may be integers only.

| Problem | Solution |
|---|---|
| a. $x^2 + 11x + 24$ | $(x + 3)(x + 8)$ |
| b. $x^2 - 11x - 12$ | $(x - 12)(x + 1)$ |
| c. $x^2 + 5x - 6$ | $(x + 6)(x - 1)$ |

**d.** $x^2 - 20x + 100$        $(x - 10)^2$

**e.** $x^2 - x - 2$        $(x - 2)(x + 1)$

**f.** $x^2 - 15x + 56$        $(x - 7)(x - 8)$

**g.** $x^2 + 8x + 16$        $(x + 4)^2$

**h.** $x^2 - 6x - 16$        $(x - 8)(x + 2)$

**i.** $x^2 - 17x + 42$        $(x - 14)(x - 3)$

**j.** $x^2 + 5xy + 6y^2$        $(x + 2y)(x + 3y)$

**k.** $3x^2 - 6x - 9$        $3(x - 3)(x + 1)$

**l.** $4x^2 + 16x - 20$        $4(x + 5)(x - 1)$

**m.** $x^3 - 12x^2 + 35x$        $x(x - 7)(x - 5)$

**n.** $2x^8 + 8x^7 + 6x^6$        $2x^6(x + 3)(x + 1)$

**o.** $x^4 - 10x^2 + 9$        $(x + 3)(x - 3)(x + 1)(x - 1)$

**p.** $x^4 - 8x^2 - 9$        $(x^2 + 1)(x + 3)(x - 3)$

**q.** $2x^2 - 5x + 3$        $(2x - 3)(x - 1)$

**r.** $2x^2 + 5x - 3$        $(2x - 1)(x + 3)$

**s.** $5x^2 - 11x + 2$        $(5x - 1)(x - 2)$

**t.** $9x^2 + 21x - 8$        $(3x + 8)(3x - 1)$

**u.** $3x^2 - 8x - 3$        $(3x + 1)(x - 3)$

**v.** $6x^2 - 13x + 6$        $(3x - 2)(2x - 3)$

**w.** $6x^2 + 35x - 6$        $(6x - 1)(x + 6)$

**x.** $9x^2 + 71x - 8$        $(9x - 1)(x + 8)$

**y.** $9x^4 + 24x^3 + 12x^2$        $3x^2(3x + 2)(x + 2)$

**Q**  **Let's do a couple of multiple-choice exercises.**

**Exercise 8:**  If $x^2 + 5x + 4 = 27$, then $3(x + 1)(x + 4) =$

A. 9                  D. 81

B. 24                 E. 243

C. 30

**Exercise 9:**  If $(x - 6)$ is a factor of $x^2 + kx - 48$, $k =$

A. −288              D. 2

B. −14               E. 14

C. −2

**A**  **Let's look at the answers.**

**Answer 8:**  D: $x^2 + 5x + 4 = (x + 1)(x + 4) = 27$; $3(x + 1)(x + 4) = 3(27) = 81$.

**Answer 9:**  D: $(-6)(?) = -48$; $? = 8$; $8 + (-6) = 2 = k$. To check, $(x - 6)(x + 8) =$ $x^2 + 2x - 48$.

Again, later on, we will present more problems involving trinomial factoring.

There are two other factorings that may occur. They are less important. However, many schools require you to know one or both of them.

## Sum and Differences of Two Cubes

We already saw that the difference of two squares factors into real numbers. For example, $x^2 - 9 = (x - 3)(x + 3)$.

The sum of two squares, such as $x^2 + 9$, does not factor into real numbers. However, with cubes, it is a different story.

$$x^3 - y^3 = (x - y)(x^2 + xy + y^2)$$

and

$$x^3 + y^3 = (x + y)(x^2 - xy + y^2).$$

**Note**  *The proof would be to multiply out the right sides and get the left side.*

**Note**  *FYI: $x^2 + 9 = (x + 3i)(x - 3i)$. We will not do this kind of factoring in this book.*

**Example 15:** Factor:

| Problem | Solution |
|---------|----------|
| **a.** $x^3 - 27$ | $(x)^3 - (3)^3 = (x - 3)(x^2 + 3x + 9)$ |
| **b.** $y^6 + 64$ | $(y^2)^3 + (4)^3 = (y^2 + 4)(y^4 - 4y^2 + 16)$ |
| **c.** $x^6 - y^6$ | $(x^3 + y^3)(x^3 - y^3) = (x + y)$ |
| | $(x^2 - xy + y^2)(x - y)(x^2 + xy + y^2)$ |

## Grouping

For an expression such as $am + em + ah + eh$:

1.   We first see there is no common factor.

2.   There is no difference of two squares since there are no squares and there are more than two terms.

3.   There is no difference or sum of two cubes since there are no cubes and there are more than two terms.

4.   There is no trinomial factoring since there are more than three terms.

Grouping the terms sometimes allows us to factor an expression. We notice in this case that an $m$ can be factored from the first two terms, and an $h$ can be factored from the last two. So $am + em + ah + eh = m(a + e) + h(a + e)$. Now we can see that the $(a + e)$ is a common factor! So the final answer is $(a + e)(m + h)$.

 **Note**   *Always try the first four types of factoring in that order. However, if grouping is possible, you may have to then use one of the first four steps.*

**Example 16:** Factor $x^3 - x^2 - x + 1$.

**Solution:**     $x^3 - x^2 - x + 1 = x^2(x - 1) - 1(x - 1)$. We have to take out $(x - 1)$ from the second two terms to make both parentheses the same! When we take out the $(x - 1)$; we get $(x - 1)(x^2 - 1) = (x + 1)(x - 1)(x - 1)$, or $(x + 1)(x - 1)^2$.

Other groupings are also possible, as shown in the following example.

**Example 17:**   Factor $x^2 - 4x + 4 - y^2$.

**Solution:**   We group this expression into three and one terms instead of two and two: $(x^2 - 4x + 4) - y^2 = (x - 2)(x - 2) - y^2 = (x - 2)^2 - y^2 = (x - 2 + y)(x - 2 - y)$, by using the difference of two squares.

With grouping, we could go on almost forever, but we won't. This is enough!

**Example 18:**   Factor the following expressions:

| Problem | Solution |
|---|---|
| a. $a^3 - 125$ | $(a - 5)(a^2 + 5a + 25)$ |
| b. $8b^3 + 343$ | $(2b + 7)(4b^2 - 14b + 49)$ |
| c. $c^6 + 27$ | $(c^2 + 3)(c^4 - 3c^2 + 9)$ |
| d. $d^6 - 1{,}000$ | $(d^2 - 10)\,(d^4 + 10d^2 + 100)$ |
| e. $8e^3 - 64$ | $8(e^3 - 8) = 8(e - 2)(e^2 + 2e + 4)$ |
| f. $f^{12} - z^{12}$ | $(f^6 + z^6)(f^6 - z^6) =$ |
| | $(f^2 + z^2)(f^4 - f^2z^2 + z^4)\,(f^3 + z^3)(f^3 - z^3) =$ |
| | $(f^2 + z^2)(f^4 - f^2z^2 + z^4) \times$ |
| | $(f + z)(f^2 - fz + z^2) \times$ |
| | $(f - z)(f^2 + fz + z^2)$ |
| g. $gx + gy + wx + wy$ | $g(x + y) + w(x + y) = (g + w)(x + y)$ |
| h. $x^3 + 3x^2 + 3x + 1$ | $x^3 + 1 + 3x^2 + 3x =$ |
| | $(x + 1)(x^2 - x + 1) + 3x(x + 1) =$ |
| | $(x + 1)(x^2 - x + 1 + 3x) =$ |
| | $(x + 1)(x^2 + 2x + 1) =$ |
| | $(x + 1)(x + 1)(x + 1) = (x + 1)^3$ |

i. $x^2 - 6xy + 9y^2 - z^2$      $(x - 3y)^2 - z^2 = (x - 3y + z)(x - 3y - z)$

j. $9 - a^2 - 2ab - b^2$      $9 - (a + b)^2 = [3 - (a + b)][3 + (a + b)] =$

$(3 - a - b)(3 + a + b)$

k. $3x - 3y + x^2 - 2xy + y^2$      $3(x - y) + (x - y)^2 = (x - y)(3 + x - y)$

(a 2 and 3 grouping)

I just put the last few in for those who want a little more challenge. If you want a super challenge, try to factor $x^4 + x^2 + 1$. Yes, it can factor. The answer is at the end of the chapter.

## ALGEBRAIC FRACTIONS

Except for adding and subtracting, the techniques for algebraic fractions are easy to understand. They must be practiced, however.

### Reducing Fractions

Factor the top and bottom; cancel factors that are the same.

**Example 19:**  Reduce the following fractions:

<u>Problem</u>

a. $\dfrac{x^2 - 9}{x^2 - 3x}$

b. $\dfrac{2x^3 + 10x^2 + 8x}{x^4 + x^3}$

c. $\dfrac{x - 9}{9 - x}$

<u>Solution</u>

$\dfrac{(x + 3)(x - 3)}{x(x - 3)} = \dfrac{x + 3}{x}$

$\dfrac{2x(x + 4)(x + 1)}{x^3(x + 1)} = \dfrac{2(x + 4)}{x^2}$

$\dfrac{(x - 9)}{-1(x - 9)} = -1$

**Ⓠ**  **Let's do a few more multiple-choice exercises.**

**Exercise 10:**  $\dfrac{16x + 4}{4} =$

A. $4x$

B. $4x + 1$

C. $12x$

D. $12x + 1$

E. $16x + 1$

**Exercise 11:**  $\dfrac{x^2 - 16}{x - 4} =$

A. $x + 4$          D. $2x - 4$

B. $x - 4$          E. $x^2 - x - 12$

C. $2x + 4$

**Exercise 12:**  $\dfrac{2x^2 + 10x + 12}{2x + 6} =$

A. $x$              D. $11x + 2$

B. $x + 1$          E. $11x + 6$

C. $x + 2$

## Ⓐ Let's look at the answers.

**Answer 10:**  B: $\dfrac{16x + 4}{4} = \dfrac{4(4x + 1)}{4} = 4x + 1.$ Another way to do this problem is

$\dfrac{16x + 4}{4} = \dfrac{16x}{4} + \dfrac{4}{4} = 4x + 1.$

**Answer 11:**  A: $\dfrac{(x + 4)(x - 4)}{x - 4} = x + 4.$

**Answer 12:**  C: $\dfrac{2(x + 2)(x + 3)}{2(x + 3)} = \dfrac{2(x + 2)}{2} = x + 2.$

## Multiplication and Division of Fractions

Algebraic fractions use the same principles as multiplication and division of numerical fractions except we factor all tops and bottoms, canceling one factor in any top with its equivalent in any bottom. In a division problem, we must remember to first invert the second fraction and then multiply.

**Example 20:**  $\dfrac{x^2 - 25}{(x + 5)^3} \times \dfrac{x^3 + x^2}{x^2 - 4x - 5} =$

**Solution:**  $\dfrac{(x + 5)(x - 5)}{(x + 5)(x + 5)(x + 5)} \times \dfrac{x^2(x + 1)}{(x - 5)(x + 1)} = \dfrac{x^2}{(x + 5)^2}$

**Example 21:**  $\dfrac{x^4 + 4x^2}{x^6} \div \dfrac{x^4 - 16}{x^2 + 3x - 10} =$

**Solution:**  $\dfrac{x^2(x^2 + 4)}{x^6} \times \dfrac{(x + 5)(x - 2)}{(x^2 + 4)(x - 2)(x + 2)} = \dfrac{x + 5}{x^4(x + 2)}$

## Adding and Subtracting Algebraic Fractions

It might be time to review the section in Chapter 2 on adding and subtracting fractions. Follow these steps:

1.  If the bottoms (denominators) are the same, add (or subtract) the tops (numerators), reducing if necessary.

2.  If the bottoms are different:

    a.  Factor the denominators of each term.

    b.  Determine the least common denominator (LCD), the product of the most number of times a prime appears in any one denominator.

    c.  Multiply top and bottom of each term by "what's missing" (i.e., the LCD divided by the denominator of that term).

    d.  Add (or subtract) and simplify the numerators; reduce if possible.

 **Note**    *The LCD is simply the least common multiple (LCM) of the denominators. LCM is discussed in Chapter 2.*

**Example 22:**  $\dfrac{x}{36 - x^2} - \dfrac{6}{36 - x^2} =$

**Solution:**  $\dfrac{x - 6}{36 - x^2} = \dfrac{x - 6}{(6 - x)(6 + x)} = \dfrac{-1}{x + 6}$

**Example 23:**  $\dfrac{5}{12xy^3} + \dfrac{9}{8x^2y} =$

**Solution:**  $\dfrac{5}{2 \times 2 \times 3xyyy} + \dfrac{9}{2 \times 2 \times 2xxy}$

$= \dfrac{5(2x)}{2 \times 2 \times 2 \times 3xxyyy} + \dfrac{9(3yy)}{2 \times 2 \times 2 \times 3xxyyy} = \dfrac{10x + 27y^2}{24x^2y^3}$

**Example 24:**  $\dfrac{2}{x^2 + 4x + 4} + \dfrac{3}{x^2 + 5x + 6} =$

**Solution:** $\dfrac{2}{(x+2)(x+2)} + \dfrac{3}{(x+2)(x+3)}$

$$= \dfrac{2(x+3)}{(x+2)(x+2)(x+3)} + \dfrac{3(x+2)}{(x+2)(x+3)(x+2)} = \dfrac{5x+12}{(x+2)(x+2)(x+3)}$$

**Example 25:** Simplify the following expressions:

| <u>Problem</u> | <u>Solution</u> |
|---|---|
| **a.** $\dfrac{x^2 - 4x + 4}{x^2 - 2x}$ | $\dfrac{x-2}{x}$ |
| **b.** $\dfrac{(x+5)^5}{x^3 + 10x^2 + 25x}$ | $\dfrac{(x+5)^3}{x}$ |
| **c.** $\dfrac{a+b-c+d-e}{-a-b+c-d+e}$ | $-1$ |
| **d.** $\dfrac{x^2 - 9}{x^2 - 9x} \times \dfrac{x^8}{x^2 + 4x + 3}$ | $\dfrac{x^7\,(x-3)}{(x-9)\,(x+1)}$ |
| **e.** $\dfrac{4x-8}{6x^2 - 6x - 12} \times \dfrac{x^2 + 2x + 1}{2x^2 + 5x + 3}$ | $\dfrac{2}{3\,(2x+3)}$ |
| **f.** $\dfrac{x^2 - y^2}{x^2 - 2xy + y^2} \times \dfrac{x^2 - 4xy + 3y^2}{x^2 - 2xy - 3y^2}$ | $1$ |
| **g.** $\dfrac{4x^2 - 9}{2x^2 - x - 3} \div \dfrac{4x^2 + 12x + 9}{x^2 + x}$ | $\dfrac{x}{2x+3}$ |
| **h.** $\dfrac{x^3 - 27}{x^3 + 1} \div \dfrac{2x^3 + 6x^2 + 18x}{x^3 - x^2 - x + 1}$ | $\dfrac{(x-3)\,(x-1)^2}{2x\,(x^2 - x + 1)}$ |
| **i.** $\dfrac{x}{9 - x^2} - \dfrac{3}{9 - x^2}$ | $\dfrac{-1}{3+x}$ |
| **j.** $\dfrac{x}{x^2 - 16} + \dfrac{4}{x+4}$ | $\dfrac{5x-16}{(x+4)\,(x-4)}$ |
| **k.** $\dfrac{3x}{x-6} - \dfrac{4}{x+2}$ | $\dfrac{3x^2 + 2x + 24}{(x-6)\,(x+2)}$ |
| **l.** $\dfrac{2}{x^2 + 5x + 6} + \dfrac{3}{x^2 + 6x + 9}$ $+ \dfrac{4}{x^2 + 7x + 12}$ | $\dfrac{9x^2 + 52x + 72}{(x+2)(x+3)(x+3)(x+4)}$ |

## SHORT AND LONG DIVISION

Algebraic **short division** is the opposite of adding fractions with like denominators.
In symbols, for example, $\dfrac{a+b-c}{d} = \dfrac{a}{d} + \dfrac{b}{d} - \dfrac{c}{d}$, where $d$ is a single term, a monomial.

**Example 26:**    Problem                                      Solution

a. $\dfrac{12x - 18}{6}$                     $2x - 3$

b. $\dfrac{8x^{10} - 12x^8 - 20x^7}{4x^4}$          $2x^6 - 3x^4 - 5x^3$

c. $\dfrac{abc - abd - ab}{ab}$              $c - d - 1$

For completeness, we will do algebraic long division. First, let's review long division with numbers.

Suppose we have $6\overline{)2,371}$. How would we do this problem?

1.  Divide 6 into 23 and get 3.

2.  $3 \times 6 = 18$.

3.  $23 - 18 = 5$.

4.  Bring down the 7 to get 57. Repeat.

5.  Divide 6 into 57 and get 9.

6.  $9 \times 6 = 54$.

7.  $57 - 54 = 3$.

8.  Bring down the 1 to get 31. Repeat.

9.  Divide 6 into 31 and get 5.

10. $5 \times 6 = 30$.

11. $31 - 30 = 1$. The remainder is 1.

12. The answer is $395\dfrac{1}{6}$.

$$
\begin{array}{r}
395\frac{1}{6} \\
6\overline{)2371} \\
\underline{18\phantom{00}} \\
57\phantom{0} \\
\underline{54\phantom{0}} \\
31 \\
\underline{30} \\
1
\end{array}
$$

> **Note**    The 6 is called the **divisor**; 2371 is called the **dividend**; $395\dfrac{1}{6}$ is called the **quotient**.

The procedure for **algebraic long division** is the same.

**Example 27:**  Divide $4x^3 - 2x^2 + 8x - 11$ by $2x - 4$

**Solution:**

$$
\begin{array}{r}
2x^2 + 3x + 10 + \dfrac{29}{2x-4} \quad x \neq 2 \\
2x - 4 \overline{)\;4x^3 - 2x^2 + 8x - 11} \\
\underline{\oplus 4x^3 \ominus 8x^2\phantom{000000000}} \\
6x^2 + 8x\phantom{0000} \\
\underline{\oplus 6x^2 \ominus 12x\phantom{0000}} \\
20x - 11 \\
\underline{\oplus 20x \ominus 40} \\
29
\end{array}
$$

1. If the divisor and/or the dividend are not arranged from the highest exponent to the lowest, rearrange the terms. In this problem, they are already arranged highest to lowest.

2. Divide the highest-power term of the divisor into the highest-power term of the dividend. $\dfrac{4x^3}{2x} = 2x^2$.

3. Multiply $2x^2$ by the entire divisor: $2x^2(2x - 4) = 4x^3 - 8x^2$.

4. Subtract: $4x^3 - 2x^2 + 80 - 11 - (4x^3 - 8x^2) = 6x^2 + 8x - 11$. Repeat.

 *We have already brought down the next term.*

 *The first terms must cancel; otherwise, there is a mistake.*

5. $\dfrac{6x^2}{2x} = 3x;\ 3x(2x - 4) = 6x^2 - 12x;\ 6x^2 + 8x - 11 - (6x^2 - 12x)$
   $= 20x - 11$. Repeat.

6. $\dfrac{20x}{2x} = 10;\ 10(2x - 4) = 20x - 40;\ 20x - 11 - (20x - 40) = 29$;
   the remainder is 29.

7. The final answer is $2x^2 + 3x + 10 + \dfrac{29}{2x - 4}$; $x \neq 2$ because you cannot divide by 0.

 *If the remainder had been 0 in this problem, that would have meant $2x - 4$ was a factor of the original polynomial.*

**Example 28:** Perform the following divisions:

| Problem | Solution |
|---|---|
| **a.** $\dfrac{20x + 5}{5}$ | $4x + 1$ |
| **b.** $\dfrac{20x^8 - 12x^5 - 4x^2}{4x^2}$ | $5x^6 - 3x^3 - 1$ |
| **c.** $\dfrac{12x^6y^8 + 10x^9y^3}{4x^2y^2}$ | $3x^4y^6 + \dfrac{5}{2}x^7y$ or $\dfrac{6x^4y^6 + 5x^7y}{2}$ |
| **d.** $\dfrac{4x^3 - 3x^2 + 2x - 1}{x + 1}$ | $4x^2 - 7x + 9 - \dfrac{10}{x + 1}$ |
| **e.** $\dfrac{4x^3 - 6x^2 + 8x - 10}{2x - 4}$ | $2x^2 + x + 6 + \dfrac{14}{2x - 4}$ |
| **f.** $\dfrac{x^3 + y^3}{x + y}$ | $x^2 - xy + y^2$ |

## Chapter 5 Quiz

1. Simplify: $4x^2 - 5y^3 - 6x^2 - 7y^3$.

2. Multiply and simplify: $4x(2x - 6) - 3x(8x - 1)$.

3. Multiply: $(3x - 4)(5x + 7)$.

4. Multiply: $4\sqrt{7}\left(3\sqrt{7} - 4 - 6\sqrt{14}\right)$.

5. $\left(x^3 - 3\sqrt{y}\right)^2 =$

6. Factor completely: $4x^4 - 64$.

7. Factor completely: $9x^2 + 21x - 8$.

8. Factor completely: $c^6 - d^6$.

9. Reduce: $\dfrac{4 - x^2}{x^2 - 4x + 4}$.

10. Add: $\dfrac{5}{x - 5} + \dfrac{2}{x^2 - 6x + 5}$.

11. Subtract: $\dfrac{4x}{x + 2} - \dfrac{2x}{x + 4}$.

12. Multiply: $\dfrac{2x + 5}{4x^2 - 25} \times \dfrac{2x - 5}{4x^2 - 20x + 25}$.

13. Divide: $\dfrac{6x - 9}{10x - 15} \div (x - 4)$.

14. Divide: $\dfrac{x^2 - 25}{x^2 - x - 20} \div \dfrac{x^2 + 8x + 15}{x^2 + 7x + 12}$.

15. Divide: $\dfrac{2x^3 - 2x^2 - 2x - 1}{2x + 2}$.

16. Divide: $\dfrac{10x^{10} - 8x^8}{2x^2}$.

17. Simplify: $\sqrt[3]{486x^6 y^{11}}$.

18. Multiply and simplify $\left(3\sqrt{2} + 2\sqrt{3}\right)\left(5\sqrt{2} + 4\sqrt{3}\right)$.

19. Simplify: $\dfrac{6x - 4}{2} - \dfrac{15x - 12}{3}$.

20. Simplify: $\dfrac{x^2 - 4x - 5}{x^2 - 7x + 10} - \dfrac{x^2 - 5x + 6}{x^2 - 2x - 3}$.

21. Rationalize the denominator: $\dfrac{\sqrt{z} + \sqrt{5}}{\sqrt{z} - \sqrt{5}}$.

22. $(3x - 7)(4x - 2) - (2x - 6)(6x + 1) =$

23. Evaluate (answer in terms of $i$): $(4 - \sqrt{-9})(2 + \sqrt{-4})$.

24. Subtract: $2 - \dfrac{x - 6}{x - 3}$.

25. Evaluate (answer in terms of $i$): $\dfrac{4 + 3i}{3 - 7i}$.

## Answers to Chapter 5 Quiz

1. $-2x^2 - 12y^3$.

2. $-16x^2 - 21x$.

3. $15x^2 + x - 28$.

4. $84 - 16\sqrt{7} - 168\sqrt{2}$.

5. $x^6 - 6x^3\sqrt{y} + 9y$.

6. $4(x^2 + 4)(x + 2)(x - 2)$.

7. $(3x + 8)(3x - 1)$.

8. $(c^3 - d^3)(c^3 + d^3) = (c - d)(c^2 + cd + d^2)(c + d)(c^2 - cd + d^2)$.

9. $\dfrac{(2 - x)(2 + x)}{(x - 2)(x - 2)} = \dfrac{-(x + 2)}{(x - 2)}$ or $\dfrac{x + 2}{2 - x}$.

10. $\dfrac{5(x - 1)}{(x - 5)(x - 1)} + \dfrac{2}{(x - 5)(x - 1)} = \dfrac{5x - 3}{(x - 5)(x - 1)}$.

11. $\dfrac{4x(x + 4) - 2x(x + 2)}{(x + 2)(x + 4)} = \dfrac{2x^2 + 12x}{(x + 2)(x + 4)}$.

12. $\dfrac{(2x + 5)}{(2x + 5)(2x - 5)} \times \dfrac{(2x - 5)}{(2x - 5)\,(2x - 5)} = \dfrac{1}{(2x - 5)\,(2x - 5)}$.

13. $\dfrac{3(2x - 3)}{5(2x - 3)} \times \dfrac{1}{(x - 4)} = \dfrac{3}{5(x - 4)}$.

14. $\dfrac{(x - 5)(x + 5)}{(x - 5)(x + 4)} \times \dfrac{(x + 4)(x + 3)}{(x + 3)(x + 5)} = 1$.

15. $x^2 - 2x + 1 + \dfrac{-3}{2x + 2}$.

16. $5x^8 - 4x^6$.

17. $\sqrt[3]{486x^6y^{11}} = \sqrt[3]{(3 \times 3 \times 3) \times (3 \times 3 \times 2) \times (xx)(xx)(xx) \times (yyy)(yyy)(yyy)(yy)}$
    $= 3x^2y^3\sqrt[3]{18y^2}$.

18. Use the F.O.I.L. method

$$\left(3\sqrt{2}\right)\left(5\sqrt{2}\right)+\left(3\sqrt{2}\right)\left(4\sqrt{3}\right)+\left(2\sqrt{3}\right)\left(5\sqrt{2}\right)+\left(2\sqrt{3}\right)\left(4\sqrt{3}\right)$$
$$=30+12\sqrt{6}+10\sqrt{6}+24=54+22\sqrt{6}.$$

19. $-2x+2.$

20. $\dfrac{(x-5)(x+1)}{(x-5)(x-2)}-\dfrac{(x-2)(x-3)}{(x+1)(x-3)}=\dfrac{(x+1)(x+1)-(x-2)(x-2)}{(x-2)(x+1)}=\dfrac{6x-3}{(x-2)(x+1)}.$

21. $\dfrac{(\sqrt{z}+\sqrt{5})(\sqrt{z}+\sqrt{5})}{(\sqrt{z}-\sqrt{5})(\sqrt{z}+\sqrt{5})}=\dfrac{z+2\sqrt{5z}+5}{z-5}.$

22. $12x^2-34x+14-(12x^2-34x-6)=20.$

23. $(4-3i)(2+2i)=8+8i-6i-6i^2=14+2i.$

24. $\dfrac{2(x-3)}{(x-3)}-\dfrac{x-6}{x-3}=\dfrac{x}{x-3}.$

25. $\dfrac{(4+3i)(3+7i)}{(3-7i)(3+7i)}=\dfrac{-9+37i}{58}=\dfrac{-9}{58}+\dfrac{37}{58}i.$

Here is the solution to the factoring of $x^4+x^2+1$. The only factors of $x^4$ are $x^2$ and $x^2$. The only factors of 1 are 1 and 1. All signs are plus. But if we multiply $(x^2+1)(x^2+1)$, we get $x^4+2x^2+1$. So we have an extra $x^2$, which has to be subtracted. So $x^4+x^2+1=(x^2+1)(x^2+1)-x^2$, or $(x^2+1)^2-x^2$, which is the difference of two squares!!!!! The answer is $(x^2+1+x)(x^2+1-x)$. For me, there are always more and more interesting problems in math (and English also). Perhaps for you also?

Let's go on to equations.

**Answer to "Bob Asks"**: Awkward.

*"Can you find a word with the letters HYX in it, in that order?"*

## FIRST-DEGREE EQUATIONS

In high school, the topic of first-degree equations was probably the most popular of all. The COMPASS asks questions that are usually not too long and usually not too tricky. To review, here are the steps to solving first-degree equations. If you get good at these, you will know when to use the steps in another order.

To solve for *x*, follow these steps (believe it or not, it took a long time to get the phrasing of this list just right):

1. Multiply by the LCD to get rid of fractions. Cross-multiply if there are only two fractions.

2. If the "*x*" term appears only on the right, switch the sides.

3. Multiply out all parentheses by using the distributive law.

4. On each side, combine like terms.

5. Add the opposite of the *x* term on the right to each side.

6. Add the opposite of the non-*x* term(s) on the left to each side.

7. Factor out the *x*. This step occurs only if there is more than one letter in a problem.

8. Divide each side by the whole coefficient of *x*, including the sign.

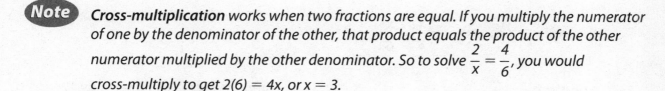

**Note**    *Cross-multiplication* works when two fractions are equal. If you multiply the numerator of one by the denominator of the other, that product equals the product of the other numerator multiplied by the other denominator. So to solve $\frac{2}{x} = \frac{4}{6}$, you would cross-multiply to get $2(6) = 4x$, or $x = 3$.

**Note**    The **opposite** of a term is the same term with its opposite sign. So the opposite of $3x$ is $-3x$, the opposite of $-7y$ is $+7y$, and the opposite of 0 is 0. The technical name for opposite is **additive inverse**.

By the x term, I mean the term with the unknown value in it that you are trying to find. It could be any other letter, but it's usually an x.

 In the following four examples, the step numbers will refer to the list of steps shown above. For each example, several numbered steps may be omitted because they are not needed to find the solution.

**Example 1:**    Solve for $x$: $7x - 2 = 10x + 13$

**Solution:**

| | |
|---|---|
| $7x - 2 = 10x + 13$ | Step 5: Add $-10x$ to each side. |
| $-3x - 2 = +13$ | Step 6: Add $+2$ to each side. |
| $-3x = 15$ | Step 8: Divide each side by $-3$. |
| $x = -5$ | Solution. |

**Example 2:**    Solve for $x$: $7 = 2(3x - 5) - 4(x - 6)$

**Solution:**

| | |
|---|---|
| $7 = 2(3x - 5) - 4(x - 6)$ | Step 2: Switch sides. |
| $2(3x - 5) - 4(x - 6) = 7$ | Step 3: Multiply out the parentheses. |
| $6x - 10 - 4x + 24 = 7$ | Step 4: Combine like terms on each side. |
| $2x + 14 = 7$ | Step 6: Add $-14$ to each side. |
| $2x = -7$ | Step 8: Divide each side by 2. |
| $x = -\dfrac{7}{2}$ | Solution. The answer doesn't have to be an integer. |

**Example 3:**    Solve for $x$: $\dfrac{x}{4} + \dfrac{x}{6} = 1$

**Solution:**

| | |
|---|---|
| $\dfrac{x}{4} + \dfrac{x}{6} = 1$ | Step 1: Multiply each term by 12. |
| $3x + 2x = 12$ | Step 4: Combine like terms. |
| $5x = 12$ | Step 8: Divide each side by 5. |
| $x = \dfrac{12}{5}$ | Solution. |

**Example 4:**    Solve for $x$: $\dfrac{4}{x-7} = \dfrac{5}{x+2}$.

**Solution:**

$\dfrac{4}{x-7} \diagup\!\!\!\!\diagup \dfrac{5}{x+2}$    Step 1: Cross-multiply.

$4(x+2) = 5(x-7)$    Step 3: Distribute.

$4x + 8 = 5x - 35$    Step 5: Add $-5x$ to each side.

$-x + 8 = -35$    Step 6: Add $-8$ to each side.

$-x = -43$    Step 8: Divide each side by $-1$.

$x = 43$    Solution

**Note**    *The answer to this problem could not be 7 or $-2$. Each would make one of the original fractions equal zero on the bottom. If one of these two numbers were the only possible answer, then the problem would have no answer because you are not allowed to divide by zero.*

**Example 5:**    Solve for $x$: $y = \dfrac{3x-5}{x-7}$

**Solution:**

Write $y = \dfrac{y}{1}$: $\dfrac{y}{1} = \dfrac{3x-5}{x-7}$    Step 1: Cross-multiply.

$(x-7)y = 1(3x-5)$    Step 3: Distribute.

$xy - 7y = 3x - 5$    Step 5: Add $-3x$ to each side.

$xy - 3x - 7y = -5$    Step 6: Add $7y$ to each side.

$xy - 3x = 7y - 5$    Step 7: Factor out the $x$ from the left.

$x(y-3) = 7y - 5$    Step 8: Divide each side by $y - 3$.

$x = \dfrac{7y-5}{y-3}$    Solution.

**Example 6:**    Solve for $x$:

| Problem | Solution |
|---|---|
| **a.** $4x - 9 = 2x - 6$ | $\dfrac{3}{2}$ |
| **b.** $4x + 5 + 6 + 7x = 0$ | $-1$ |
| **c.** $4(5x - 6) = 7(8x - 9)$ | $\dfrac{13}{12}$ |
| **d.** $10(2x - 3) - 30x = 45$ | $-7.5$ |
| **e.** $ax + b = c$ | $\dfrac{c-b}{a}$ |

**f.** $\dfrac{1}{x} + \dfrac{1}{5} = \dfrac{1}{2}$              $\dfrac{10}{3}$

**g.** $y = \dfrac{3}{x - 5}$              $\dfrac{3 + 5y}{y}$

**h.** $(x + 5)(y + z) = 6$              $\dfrac{6 - 5y - 5z}{y + z}$

**i.** $(x + 2)(x + 3) - x(x + 8) = 24$   $-6$

**j.** $7x = 10x$                              $0$

**Q** **Let's do some multiple-choice exercises.**

**Exercise 1:**    $2x - 6 = 4; x + 3 =$

A. 5                    D. 8

B. 6                    E. 9

C. 7

**Exercise 2:**    $x - 9 = 9 - x; x =$

A. 0                    D. 13.5

B. 4.5                 E. 18

C. 9

**Exercise 3:**    $4x - 17 = 32; 12x - 51 =$

A. $12\dfrac{1}{4}$              D. 96

B. $36\dfrac{3}{4}$              E. 288

C. 64

**Exercise 4:**    $\dfrac{xy}{y - x} = 1; x =$

A. $\dfrac{1}{2}$                  D. $\dfrac{y}{y - 1}$

B. 1                    E. $\dfrac{y}{1 - y}$

C. $\dfrac{y}{y + 1}$

**Let's look at the answers.**

**Answer 1:**    D: $x = 5$, but the question asks for $x + 3 = 8$.

**Answer 2:**    C: $2x = 18$; $x = 9$. The answer is C. It actually can be solved just by looking because $9 - 9 = 9 - 9$ (or $0 = 0$).

**Answer 3:**    D: We do not solve this equation. We recognize that $12x - 51 = 3(4x - 17) = 3(32) = 96$.

**Answer 4:**    C: By cross-multiplying, we get $xy = y - x$. Then $xy + x = y$, which factors to $x(y + 1) = y$. Therefore, $x = \dfrac{y}{y + 1}$.

# LINEAR INEQUALITIES

To review some facts about inequalities:

$a < b$ (read, "$a$ is less than $b$") means $a$ is to the left of $b$ on the number line.

$a > b$ (read, "$a$ is greater than $b$") means $a$ is to the right of $b$ on the number line.

$x > y$ is the same as $y < x$.

The notation $x \geq y$ (read, "$x$ is greater than or equal to $y$") means $x > y$ or $x = y$.

Similarly, $x \leq y$ (read, "$x$ is less than or equal to $y$"), means $x < y$ or $x = y$.

We solve linear inequalities ($<$, $>$, $\leq$, $\geq$) the same way we solve linear equalities, except when we multiply or divide by a negative, the inequality sign reverses direction.

**Example 7:**    Solve for $x$: $6x + 2 < 3x + 10$

**Solution:**    $3x < 8$, so $x < \dfrac{8}{3}$. The inequality does not switch because both sides are divided by a positive number (3).

**Example 8:**    Solve for $x$: $-2(x - 3) \leq 4x - 3 - 7$

**Solution:**    $-2x + 6 \leq 4x - 10$, or $-6x \leq -16$. Thus, $x \geq \dfrac{-16}{-6} = \dfrac{8}{3}$. Here the inequality switches because we divided both sides by a negative number ($-6$).

**Example 9:**    Solve for $x$: $8 > \dfrac{x-2}{-3} \geq 5$

**Solution:**    We multiply through by $-3$, and both inequalities switch. We get $-24 < x - 2 \leq -15$. If we add 2 to each part to get a value for $x$ alone, the final answer is $-22 < x \leq -13$.

**Note**    *This answer is read as "x is greater than $-22$ and less than or equal to $-13$." This type of "double" inequality occurs when x is between two values.*

**Note**    *Sometimes the x value can be outside of two values, such as $x < -3$ or $x > 2$, read as "x is less than $-3$ or greater than 2." Use "or" because x cannot be both at the same time.*

**Example 10:**    Solve for $x$:

| Problem | Solution |
|---|---|
| **a.** $10x - 5 < 5x - 11$ | $x < -\dfrac{6}{5}$ |
| **b.** $-2x - 7 < 4x + 9$ | $x > -\dfrac{8}{3}$ |
| **c.** $4(2x - 3) - 7x \leq 12 - 2x$ | $x \leq 8$ |
| **d.** $1 - 2x - 7 \leq -6$ | $x \geq 0$ |
| **e.** $1 \leq 2(3x - 5) < 7$ | $\dfrac{11}{6} \leq x < \dfrac{17}{6}$ |
| **f.** $1 < \dfrac{2x-5}{-3} \leq 7$ | $1 > x \geq -8$ or $-8 \leq x < 1$ |

**Q**    **Now, let's do some more exercises.**

**Exercise 5:**    If $3x + 4 > 17$, $3x + 7 >$

A. $\dfrac{13}{3}$            D. 17

B. $\dfrac{22}{3}$            E. 20

C. 14

**Exercise 6:**    If $3x + 4y < 5$, $x <$

A. $5 - 4y - 3$            D. $\dfrac{5 - 4y}{3}$

B. $\dfrac{5}{4}y - 3$            E. $\dfrac{5}{3} - 4y$

C. $\dfrac{5}{12}y$

**Exercise 7:** $x > 0$ and $y > 0$. The number of ordered pairs of whole numbers $(x, y)$ such that $2x + 3y < 9$ is

**A.** 1 **D.** 4

**B.** 2 **E.** 5

**C.** 3

(A) **Let's look at the answers.**

**Answer 5:** **E:** We don't actually have to solve this one. $3x + 7 = (3x + 4) + 3 > 17 + 3$, or 20.

**Answer 6:** **D:** $3x + 4y < 5$ is the same as $3x < 5 - 4y$. Dividing by 3, we get $\frac{5 - 4y}{3}$.

**Answer 7:** **C:** We must substitute numbers. (1, 1) is okay because $2(1) + 3(1) < 9$; (2, 1) is okay because $2(2) + 3(1) < 9$; (1, 2) is okay because $2(1) + 3(2) < 9$; and that's all.

We will do more on ordered pairs later in the book. As we have seen already, some questions overlap more than one topic.

## ABSOLUTE VALUE EQUALITIES

**Absolute value** is the magnitude of a number, without regard to sign. You should know the following facts about absolute value:

$|3| = 3, |-7| = 7$, and $|0| = 0$

$|u| = 6$ means that $u = 6$ or $-6$.

$|u| = 0$ always means $u = 0$.

$|u| = -17$ has no solutions because the absolute value is never negative.

**Example 11:** Solve for $x$: $|2x - 5| = 7$

**Solution:** Either $2x - 5 = 7$ or $2x - 5 = -7$. So $x = 6$ or $x = -1$.

(Note) *This kind of problem always has two answers.*

**Example 12:** Solve for $x$: $|5x + 11| = 0$.

**Solution:** $5x + 11 = 0$; $x = -\frac{11}{5}$.

(Note) *This type of problem (absolute value equals 0) always has one answer.*

**Example 13:** Solve for $x$:

| Problem | Solution |
|---|---|
| **a.** $|3x - 5| = 7$ | $x = 4, -\dfrac{2}{3}$ |
| **b.** $|5 - 3x| = 7$ | $x = 4, -\dfrac{2}{3}$ |
| **c.** $|4x + 7| = 3$ | $x = -1, -\dfrac{5}{2}$ |
| **d.** $|7x + 3| = 0$ | $x = -\dfrac{3}{7}$ |
| **e.** $|7x - 2| = -2$ | No answer; the absolute value cannot be negative. |

**Q** **Let's do some more multiple-choice exercises:**

**Exercise 8:** $|2x + 1| = |x + 5|$ ; $x =$

A. $-2$          D. 4 and $-2$

B. 0             E. 0 and 4

C. 4

**Exercise 9:** $|x - y| = |y - x|$. This statement is true

A. for no values          D. only for all integers

B. only if $x = y = 0$    E. for all real numbers

C. only if $x = y$

**Exercise 10:** If $4|2x + 3| = 11$, $8|2x + 3| + 5 =$

A. $\dfrac{11}{4}$          D. 27

B. $\dfrac{31}{4}$          E. 110

C. 22

**A** **Let's look at the answers.**

**Answer 8:** D: $2x + 1 = x + 5$ or $2x + 1 = -(x + 5)$, for which the answers are 4 and $-2$, respectively.

**Answer 9:**  E: For example, if $x = 3$ and $y = -2$, $|3 - (-2)| = |-2-3| = 5$.

**Answer 10:**  D: We don't have to solve this for $x$ at all. If $4|2x + 3| = 11$, then $8|2x + 3| = 2(11) = 22$. Adding 5, we get 27.

# QUADRATIC EQUATIONS

**Quadratic equations**, equations involving the square of the variable, can be solved in three principal ways: factoring, taking the square root, and using the quadratic formula. Another name for the solution of any equation is a **root**. An equation has as many roots as the highest power of the unknown variable.

## Solving Quadratics by Factoring

Solving quadratic equations by factoring is based on the fact that if $a \times b = 0$, then either $a = 0$ or $b = 0$.

**Example 14:**  Solve for all values of $x$: $x(x - 3)(x + 7)(2x + 1)(3x - 5)(ax + b)(cx - d) = 0$.

**Solution:**  Setting each factor equal to 0 (better if you can do it just by looking), we get $x = 0, 3, -7, -\dfrac{1}{2}, \dfrac{5}{3}, -\dfrac{b}{a}$, and $\dfrac{d}{c}$.

## Solving Quadratics by Taking the Square Root

If the equation is of the form $x^2 = c$, with no $x$ term, we just take the square root: $x = \pm\sqrt{c}$.

 *Remember that $\sqrt{9} = 3$, $-\sqrt{9} = -3$, $\sqrt{-9}$ is not real, and if $x^2 = 9$, then $x = \pm 3$!*

**Example 15:**  Solve for all values of $x$:

    **a.** $x^2 - 7 = 0$

    **b.** $ax^2 - b = c$, where $a, b, c > 0$.

**Solutions:**  **a.** $x^2 = 7$; $x = \pm\sqrt{7}$

    **b.** $ax^2 = b + c$; $x^2 = \dfrac{b + c}{a}$, so $x = \pm\sqrt{\dfrac{b + c}{a}}$, or $\dfrac{\pm\sqrt{a(b + c)}}{a}$

**Example 16:**  Solve for $x$ by factoring:

| Problem | Solution |
|---|---|
| **a.** $x^2 - 5x - 6 = 0$ | $x = 6, -1$ |
| **b.** $x^2 - 3x - 40 = 0$ | $x = 8, -5$ |

   **c.** $3x^2 - 10x + 3 = 0$ $\qquad\qquad$ $x = \frac{1}{3}, 3$

   **d.** $x^3 + 7x^2 - 8x = 0$ $\qquad\qquad$ $x = 1, 0, -8$

   **e.** $x^2 - 7x + 9 = 3$ $\qquad\qquad$ $x = 1, 6$

   **f.** $2x^3 - 18x = 0$ $\qquad\qquad$ $x = 3, 0, -3$

   **g.** $x^4 - 10x^2 + 9 = 0$ $\qquad\qquad$ $x = \pm 1, \pm 3$

   **h.** $x^4 - 8x^2 - 9 = 0$ $\qquad\qquad$ $x = \pm 3$

**Example 17:** Solve for $x$ by taking the square root:

   **a.** $x^2 = 9$ $\qquad\qquad$ $x = \pm\sqrt{9} = \pm 3$

   **b.** $x^2 = 11$ $\qquad\qquad$ $x = \pm\sqrt{11}$

   **c.** $5x^2 - 16 = 0$ $\qquad\qquad$ $x = \dfrac{\pm 4\sqrt{5}}{5}$

   **d.** $Ax^2 + M + I = T$ $\qquad\qquad$ $x = \pm\dfrac{\sqrt{A(T - I - M)}}{A}$

   **e.** $(mx + n)^2 = p$ $\qquad\qquad$ $x = \dfrac{\pm\sqrt{p} - n}{m}$

**Q** **Let's do some more multiple-choice exercises:**

**Exercise 11:** The solutions to $x^2 - 3x - 4 = -6$ are

   **A.** $x = 4$ and $x = -1$ $\qquad$ **D.** $x = 1$ and $x = 2$

   **B.** $x = 4$ and $x = 1$ $\qquad$ **E.** $x = 5$ and $x = -2$

   **C.** $x = 1$ and $x = -1$

**Exercise 12:** If $x = 6$ and $x = -9$ are the answers to a quadratic equation, the original quadratic equation is

   **A.** $x^2 + 15x - 54 = 0$ $\qquad$ **D.** $x^2 + 3x - 54 = 0$

   **B.** $x^2 - 15x - 54 = 0$ $\qquad$ **E.** $x^2 - 54x + 54 = 0$

   **C.** $x^2 - 3x - 54 = 0$

**Exercise 13:** $x = 2a$ and $x = -5b$ are solutions to a quadratic equation. The quadratic equation is

A. $x^2 - 2ab - 10ab = 0$    D. $x^2 - (2a - 5b)x - 10ab = 0$

B. $x^2 + 2ab - 10ab = 0$    E. $x^2 + (2x - 5b)x + 10ab = 0$

C. $x^2 + (2a - 5b)x - 10ab = 0$

 **Let's look at the answers.**

**Answer 11:**    D: Add 6 to both sides and factor to get $x^2 - 3x + 2 = (x - 2)(x - 1) = 0$. Then set each factor equal to 0.

**Answer 12:**    D: If 6 is a solution, $(x - 6)$ is a factor, and if $-9$ is a solution, $(x + 9)$ is a factor. So $(x - 6)(x + 9) = x^2 + 3x - 54 = 0$.

**Answer 13:**    D: $(x - 2a)(x + 5b) = x^2 - 2ax + 5bx - 10ab = x^2 - (2a - 5b)x - 10ab = 0$.

## Solving Quadratics by Using the Quadratic Formula

The quadratic formula states that if $ax^2 + bx + c = 0$, then

$$x = \frac{-b \pm \sqrt{b^2 - 4ac}}{2a},$$

where $a$ is the coefficient of the $x^2$ term, $b$ is the coefficient of the $x$ term, and $c$ is the number term.

The COMPASS sometimes asks you to find the sum or the product of the roots. You can do this without actually finding the roots. If $ax^2 + bx + c = 0$, then the sum of the roots is $-\frac{b}{a}$ and the product of the roots is $\frac{c}{a}$. As an example, if $2x^2 + 3x - 6 = 0$, the sum of the roots is $-\frac{3}{2}$ and the product of the roots is $\frac{-6}{2} = -3$.

**Example 18:**    Solve $3x^2 - 5x + 2 = 0$ by using the quadratic formula.

**Solution:**    $a = 3, b = -5, c = 2$. $x = \dfrac{-(-5) \pm \sqrt{(-5)^2 - 4(3)(2)}}{2(3)} = \dfrac{5 \pm 1}{6}$. $x_1$ (read as "x sub one," the first answer) $= \dfrac{5 + 1}{6} = 1$; $x_2$ (read, "x sub two," the second answer) $= \dfrac{5 - 1}{6} = \dfrac{2}{3}$.

**Example 19:**  Solve $3x^2 - 5x + 2 = 0$ by factoring.

**Solution:**  This is the same problem as Example 18. $3x^2 - 5x + 2 = (3x - 2)(x - 1)$ $= 0$, so $x = \dfrac{2}{3}$ or 1.

Factoring is preferred; using the quadratic formula takes too long.

**Example 20:**  Find the sum and product of the roots of $5x^2 + 7x - 11 = 0$.

**Solution:**  In this example, $a = 5$, $b = 7$, and $c = -11$. Thus, the sum of the roots is $-\dfrac{7}{5}$ and the product of the roots is $-\dfrac{11}{5}$.

Before we go to the next set of exercises, I suppose most of you know the quadratic formula, but few have seen it shown to be true. The teacher in me has to show you, even if you don't care.

| | |
|---|---|
| $ax^2 + bx + c = 0$ | The coefficient of $x^2$ must be 1. |
| $x^2 + \dfrac{b}{a}x = -\dfrac{c}{a}$ | Complete the square; this means taking half the coefficient of $x$, squaring it, and adding it to both sides. |
| $x^2 + \dfrac{b}{a}x + \left(\dfrac{b}{2a}\right)^2 = \left(\dfrac{b}{2a}\right)^2 - \dfrac{c}{a}$ | Factor the left side and do the algebra on the right side. |
| $\left(x + \dfrac{b}{2a}\right)^2 = \dfrac{b^2}{4a^2} - \dfrac{c\,(4a)}{a\,(4a)} = \dfrac{b^2 - 4ac}{4a^2}$ | Multiply the last term by $\dfrac{4a}{4a}$ and combine terms. |
| $\left(x + \dfrac{b}{2a}\right)^2 = \dfrac{b^2}{4a^2} - \dfrac{4ac}{4a^2} = \dfrac{b^2 - 4ac}{4a^2}$ | Take the square root of both sides. |
| $x + \dfrac{b}{2a} = \dfrac{\pm\sqrt{b^2 - 4ac}}{2a}$ | Solve for $x$ and simplify. |
| $x = -\dfrac{b}{2a} \pm \dfrac{\sqrt{b^2 - 4ac}}{2a} = \dfrac{-b \pm \sqrt{b^2 - 4ac}}{2a}$ | |

The quadratic formula is really true!

We now know for sure that if $ax^2 + bx + c = 0$, then $x = \dfrac{-b \pm \sqrt{b^2 - 4ac}}{2a}$, where $a$ is the coefficient of $x^2$, $b$ is the coefficient of $x$, and $c$ is the non-$x$ term. Let's make sure you really understand this.

**Example 21:** For each problem, tell what $a$ is, what $b$ is, and what $c$ is.

Problem

Solution

| | $a$ | $b$ | $c$ |
|---|---|---|---|
| **a.** $3x^2 - 4x - 7 = 0$ | 3 | $-4$ | $-7$ |
| **b.** $x^2 - 15 = 0$ | 1 | 0 | $-15$ |
| **c.** $x^2 + 2x + \dfrac{7}{4} = 0$ | 1 | 2 | $\dfrac{7}{4}$ |
| **d.** $3x - x^2 = 0$ | $-1$ | 3 | 0 |

(Rewrite as $-x^2 + 3x = 0$.)

**e.** $a^2x^2 - bx - c = 0$     $a^2$     $-b$     $-c$

(The coefficient of $x^2 = a^2$; in front of $x$ is $-b$; $-c$ (not its coefficient) is the term without $x$!)

**f.** $-x^2 + b^2 + 7 = 0$     $-1$     0     $b^2 + 7$

(No $x$ term; two terms without $x$.)

**g.** $3x^2 + ax + 4x - 2 = 0$     3     $a + 4$     $-2$

(Rewrite $ax + 4x$ as $(a + 4)x$; coefficient of $x$ is $(a + 4)$.)

**h.** $y = x^2 - 2x + 3$     1     $-2$     $3 - y$

(Rewrite as $x^2 - 2x + (3 - y) = 0$.)

**Example 22:** For each of the problems in Example 22, solve for $x$ by using the quadratic formula.

Problem

Solution

**a.** $3x^2 - 4x - 7 = 0$

$$x = \frac{-(-4) \pm \sqrt{(-4)^2 - 4(3)(-7)}}{2(3)} = -1, \frac{7}{3}$$

**b.** $x^2 - 15 = 0$

$$x = \frac{0 \pm \sqrt{0^2 - 4(1)(-15)}}{2(1)} = \pm\sqrt{15}$$

**c.** $3x - x^2 = 0$

$$x = \frac{-3 \pm \sqrt{3^2 - 4(-1)(0)}}{2(-1)} = 0, 3$$

(On a test, you would solve this one by factoring.)

**d.** $a^2x^2 - bx - c = 0$

$$x = \frac{-(-b) \pm \sqrt{(-b)^2 - 4(a^2)(-c)}}{2a^2}$$

$$= \frac{b \pm \sqrt{b^2 + 4a^2c}}{2a^2}$$

(Ugly, but correct.)

**e.** $-x^2 + b^2 + 7 = 0$

$$x = \frac{0 \pm \sqrt{0^2 - 4(-1)(b^2 + 7)}}{2(-1)} = \pm\sqrt{b^2 + 7}$$

(Ugly again; the higher you go in math, the uglier the answers can get (not always, but more often!))

**f.** $3x^2 + ax + 4x - 2 = 0$

$$\frac{-(a + 4) \pm \sqrt{(a + 4)^2 - 4(3)(-2)}}{2(3)}$$

$$= \frac{-a - 4 \pm \sqrt{a^2 + 8a + 40}}{6}$$

**g.** $y = x^2 - 2x + 3$

$$x = \frac{-(-2) \pm \sqrt{(-2)^2 - 4(1)(3 - y)}}{2(1)}$$

$$= \frac{2 \pm \sqrt{4y - 8}}{2} = 1 \pm \sqrt{y - 2}$$

(This last problem is used to find the inverse map with functions. The values of $x$ determine whether we use the $+$ or $-$ sign. We will have more about this later.)

Now, let's do the chapter review test.

## Chapter 6 Quiz

In each question, solve for $x$.

1. $3x - 9 = -7x + 33$.

2. $4(2x - 3) - 5(4x - 1) = 6$.

3. $ax + by = c$.

4. $y = \dfrac{3x - 5}{x - 2}$.

5. $\dfrac{4}{x - 7} = \dfrac{7}{x - 4}$.

6. $7x^2 - 19 = 0$.

7. $5x^2 - x - 10 = 0$.

8. $|3x - 7| = 5$.

9. $4x - 6 < 7x + 24$.

10. $|4x + 7| = -3$.

11. $x^4 - 26x^2 + 25 = 0$.

12. What is the sum of the roots of $5x^2 - 4x - 7 = 0$?

13. What is the product of the roots of $5x^2 - 4x - 7 = 0$?

14. Given the equation $a^2x^2 - bx + c = 0$, what is the expression for the value of $x$?

15. $cx^2 - bx + a = 0$.

16. $|2x - 5| = |4x - 1|$.

17. $-4 < -2x - 6 \le 10$.

18. $m(x - y) + r = s$.

19. $x^3 - 20x^2 = 21x$.

20. $9x = 11x$.

**Answers to Chapter 6 Quiz**

1. 4.2

2. $-\dfrac{13}{12}$.

3. $\dfrac{c - by}{a}$.

4. $\dfrac{2y - 5}{y - 3}$.

5. 11.

6. $\dfrac{\pm\sqrt{133}}{7}$.

7. $\dfrac{1 \pm \sqrt{201}}{10}$.

8. $4, \dfrac{2}{3}$.

9. $x > -10$.

10. No solution! The absolute value can never equal a negative number.

11. $\pm 1, \pm 5.$

12. $-\dfrac{b}{a} = -\left(\dfrac{-4}{5}\right) = \dfrac{4}{5}.$

13. $\dfrac{c}{a} = -\dfrac{7}{5}.$

14. $a = a^2, b = -b,$ and $c = c,$ so $x = \dfrac{b \pm \sqrt{b^2 - 4a^2c}}{2a^2}.$

15. $\dfrac{b \pm \sqrt{b^2 - 4ac}}{2c}.$

16. $-2, 1.$

17. $-8 \leq x < -1.$

18. $\dfrac{s - r + my}{m}.$

19. $-1, 0, 21.$

20. $0.$

Let's go on.

**Answer to "Bob Asks":** Asphyxiate.

# CHAPTER 7:  *Words and Word Problems*

*"What word contains three double letters in a row?"*

**In** a certain way, this was the hardest chapter to write. This area is very necessary and is severely neglected in high school. The placement test may not ask very many word problems other than the ones we have seen for arithmetic. Yet if you go on in math, not seeing these problems will be very costly to you. I decided to include here the most likely ones you might see. If you want more, go to my *Algebra* book for the ones you didn't see, my *Precalculus* book for a preview for calculus, my *Calc 1* and *Calc 3* books for those type of word problems, my *GRE* book when you go to a general graduate school, and my *GMAT* book when going for your MBA. These books are listed in the beginning of this book.

Too much talking! Let's get started.

## BASICS

As we know, the answer in **addition** is the **sum**. Other words that indicate addition are **plus**, **more**, **more than**, **increase**, and **increased by**. You can write all sums in any order because addition is commutative.

The answer in **multiplication** is the **product**. Another word that is used is **times**. Sometimes the word **of** indicates multiplication, as we shall see shortly. **Double** means to multiply by two, and **triple** means to multiply by three. Because multiplication is also commutative, we can write any product in any order.

Division's answer is called the **quotient**. Another phrase that is used is **divided by**.

The answer in **subtraction** is called the **difference**. Subtraction can present a reading problem because $4 - 6 \neq 6 - 4$, so we must be careful to subtract in the correct order. Example 1 shows how some subtraction phrases are translated into algebraic expressions.

**Example 1:**    <u>Phrases</u>                                              <u>Expressions</u>

    **a.** The difference between 9 and 5                    $9 - 5$

         The difference between $m$ and $n$              $m - n$

    **b.** Five minus two                                    $5 - 2$

         $m$ minus $n$                                    $m - n$

    **c.** Seven decreased by three                         $7 - 3$

         $m$ decreased by $n$                             $m - n$

    **d.** Nine diminished by four                          $9 - 4$

         $m$ diminished by $n$                            $m - n$

    **e.** Ten less two                                     $10 - 2$

         $m$ less $n$                                     $m - n$

    **f.** Ten less than two                                $2 - 10$

         $m$ less than $n$                                $n - m$

    **g.** Three from five                                  $5 - 3$

         $m$ from $n$                                     $n - m$

**Note**    *One word can make a big difference. In parts e and f of Example 1, m less n means m−n; and m less than n means n − m. In addition, m is less than n means m < n. You must read carefully!*

The following words usually indicate an equal sign: *is, am, are, was, were, the same as, equal to.*

You also must know the following phrases for inequalities: at least ($\geq$), not more than ($\leq$), over ($>$), and under ($<$).

**Example 2:**    Write the following in symbols:

| Problem | Solution |
|---|---|
| **a.** $m$ times the sum of $q$ and $r$ | $m(q + r)$ |
| **b.** Six less the product of $x$ and $y$ | $6 - xy$ |
| **c.** The difference between $c$ and $d$ divided by $f$ | $\dfrac{c - d}{f}$ |
| **d.** $b$ less than the quotient of $r$ divided by $s$ | $\dfrac{r}{s} - b$ |
| **e.** The sum of $d$ and $g$ is the same as the product of $h$ and $r$ | $d + g = hr$ |
| **f.** $x$ is at least $y$ | $x \geq y$ |
| **g.** Zeb's age $n$ is not more than 21 | $n \leq 21$ |
| **h.** I am over 30 years old, where "$I$" represents "my age" | $I > 30$ |
| **i.** Most people are under seven feet tall, where "$p$" represents "most people" | $p < 7$ |

## COMMON AND EVERYDAY PROBLEMS

**Example 3:**    In spring at a recent school board election, Ms. Johnson received 50 more votes than Ms. Smith for president of the school board. If 892 votes were cast, how many votes did each get?

**Solution:**    In most problems, it is better to let $x$ = the smaller number—in this case, Ms. Smith's vote total. Ms. Johnson received $x + 50$ votes. Then $x + (x + 50) = 892$; $2x = 842$; so $x = 421$, the number of Ms. Smith's votes. Ms. Johnson received $421 + 50 = 471$ votes.

**Example 4:**    Two numbers total 40. The sum of twice one of them and triple the other is 105. Find the numbers.

**Note**    *If two numbers total 40 and one is 11, the other is 40 − 11 (or 29). If two numbers total 40 and one is unknown, x, the other is 40 − x.*

**Solution:**    Let one number = $x$; the other number is $40 - x$. The sum (add) of twice one number, $2x$, and triple the other number, $3(40 - x)$, is 105. The equation is $2x + 3(40 - x) = 105$; $2x + 120 - 3x = 105$; $x = 15$ and, because both numbers were asked for, $40 - x = 40 - 15 = 25$ is the other number.

**Example 5:**    The sum of two numbers is 20. Five times the larger less the smaller is 70. Find the larger number.

**Solution:**    Let $x$ = the number you are looking for—in this case, the larger number. Then $20 - x$ is the smaller number. *Less* means "subtract," so don't change the order. The equation is $5x - (20 - x) = 70$; $6x - 20 = 70$; $x = 15$.

## RATIOS AND PROPORTIONS

The comparison of two numbers is called a **ratio**. The ratio of 3 to 5 is written two ways: $\dfrac{3}{5}$ or 3 : 5 (read, "the ratio of 3 to 5").

**Example 6:**    Find the ratio of 5 ounces to 2 pounds.

**Solution:**    The ratio is $\dfrac{5}{32}$, because 16 ounces are in a pound, and a ratio should have the same measurements.

**Example 7:**    A board is cut into two pieces that are in the ratio of 3 to 4. If the board is 56 inches long, how long is the longer piece?

**Solution:**    If the pieces are in the ratio 3 : 4, we let one piece equal $3x$ and the other $4x$. The equation, then, is $3x + 4x = 56$; so $x = 8$; and the longer piece is $4x = 32$ inches.

I have asked this problem many, many times. Almost no one has ever gotten it correct—not because it is difficult, but because no one does problems like this anymore.

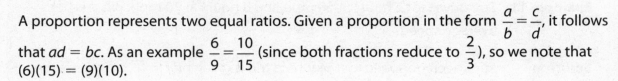

A proportion represents two equal ratios. Given a proportion in the form $\dfrac{a}{b} = \dfrac{c}{d}$, it follows that $ad = bc$. As an example $\dfrac{6}{9} = \dfrac{10}{15}$ (since both fractions reduce to $\dfrac{2}{3}$), so we note that $(6)(15) = (9)(10)$.

**Example 8:** Solve for $x$: $\dfrac{x-3}{4} = \dfrac{5}{7}$.

**Solution:** Cross-multiplying, we get $(7)(x - 3) = (4)(5)$. This equation simplifies to $7x - 21 = 20$. Then $7x = 41$, so $x = \dfrac{41}{7}$ or $5\dfrac{6}{7}$.

**Example 9:** On a map, 2 inches represents a distance of 45 miles. How many miles on this map are represented by 9 inches?

**Solution:** Set up a proportion in which the number of inches is placed in the numerators and the number of miles is placed in the denominators. Let $x$ represent the unknown number of miles. Then $\dfrac{2}{45} = \dfrac{9}{x}$, which becomes $2x = (45)(9) = 405$.

Thus, $x = \dfrac{405}{2} = 202.5$.

**Example 10:** If $a$ oranges cost $b$ cents, how many oranges can you buy for $d$ dollars?

**Solution:** Set up a proportion in which the number of oranges is placed in the numerators and the cost in cents is placed in the denominators. Let $x$ represent the required number of oranges. We must change $d$ dollars to $100d$ cents. Then $\dfrac{a}{b} = \dfrac{x}{100d}$, which becomes $100ad = bx$. Now divide both sides by $b$ to get $x = \dfrac{100ad}{b}$.

## SPEED

We are familiar with speed being given in miles per hour (mph), so it is easy to remember that

$\text{speed} = \dfrac{\text{distance}}{\text{time}}$, or $r = \dfrac{d}{t}$, where $r$ stands for rate (the speed). Use this relationship, or the equivalent ones: $d = rt$ or $t = \dfrac{d}{r}$, to do word problems involving speed.

**Example 11:** Sue drives for 2 hours at 60 mph and 3 hours at 70 mph. What is her average speed?

**Solution:** Sue's average speed for the whole trip is given by $r = \dfrac{d}{t}$, where $d$ is the total distance and $t$ is the total time. Use $d = rt$ for each part of her trip to get the total distance. The total distance is $60(2) + 70(3) = 330$ miles. The total time is 5 hours. So Sue's average speed is $r = \dfrac{330}{5}$ mph = 66 mph.

**Note** *The average speed is not the simple mean of two or more speeds. It is the total distance divided by the total time.*

**Example 12:** Don goes 40 mph in one direction and returns at 60 mph. What is his average speed?

**Solution:** Notice that the problem doesn't tell the distance. It doesn't have to; the distance in each direction is the same, because it is a round trip. We can take any distance, so let's choose 120 miles, the LCM of 40 and 60. Then the time going is $\dfrac{120}{40} = 3$ hours, and the time returning is $\dfrac{120}{60} = 2$ hours. The average speed is the total distance divided by the total time, $\dfrac{2(120)}{3+2} = \dfrac{240}{5} = 48$ mph.

We actually don't have to choose a number for the distance, however. We could use $x$. Just for learning's sake, we will do this same problem (Example 12) using $x$ as the distance. The time going is $\dfrac{x}{40}$, and the time returning is $\dfrac{x}{60}$. Then we have:

$$\text{Total speed} = \frac{\text{total distance}}{\text{total time}} = \frac{2x}{\frac{x}{40} + \frac{x}{60}} = \frac{120(2x)}{120\left(\frac{x}{40} + \frac{x}{60}\right)} = \frac{240x}{5x} = 48 \text{ mph}$$

**Example 13:**   A plane leaves Indianapolis traveling west. At the same time, a plane traveling 30 mph faster leaves Indianapolis going east. After two hours the planes are 2,000 miles apart. What is the speed of the faster plane?

**Solution:**   A chart and a picture are best for problems like this.

|   | $r$ | $t$ | $d$ |
|---|-----|-----|-----|
| W | $x$ | 2 | $2x$ |
| E | $x + 30$ | 2 | $2(x + 30)$ |

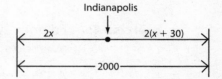

We let $x =$ the speed of the plane going west; then $x + 30$ is the speed of the eastbound plane. The time for each is 2 hours. Because $rt = d$, the distances are as shown in the above chart. According to the picture, $2x + 2(x + 30) = 2000$. So $x = 485$; $x + 30 = 515$ mph.

The problem is the same if the planes are starting at the ends and flying toward each other.

 **Let's do some basic multiple-choice exercises.**

**Exercise 1:**   Four more than a number is seven less than triple the number. The number is

A. 4

B. 5.5

C. 7

D. 9

E. 11

**Note**   *The word "number" does not necessarily mean an integer or even necessarily a positive number.*

**Exercise 2:**   Mike must have at least an 80 average but less than a 90 average to get a *B*. If he received scores of 98, 92, and 75 on the first three tests, which of these grades on the fourth test will give him a *B*?

A. 42

B. 54

C. 66

D. 98

E. 100

**Exercise 3:**    Nine less than a number is the same as the difference between nine and the number. The number is

A. 18    D. −9

B. 9    E. All numbers are correct.

C. 0

**Exercise 4:**    Ed goes 20 mph in one direction and 50 mph on the return trip. His average speed is

A. 25 mph    D. 30 mph

B. $27\dfrac{2}{7}$ mph    E. $30\dfrac{6}{7}$ mph

C. $28\dfrac{4}{7}$ mph

**Exercise 5:**    The angles of a triangle are in the ratio of 3 : 5 : 7. The largest angle is

A. 12°    D. 84°

B. 36°    E. 108°

C. 60°

**Exercise 6:**    The value of $d$ dimes and $q$ quarters in pennies is

A. $d + q$    D. $10d + 25q$

B. $dq$    E. $35dq$

C. $250dq$

**Exercise 7:**    A car leaves Chicago at 2 P.M. going west. A second car leaves Chicago at 5 P.M., going west at 30 mph faster. At 7 P.M., the faster car hits the slower one. The accident occurred after how many miles?

A. 200    D. 40

B. 100    E. 20

C. 60

**Exercise 8:** A fraction, when reduced, is $\frac{2}{3}$. If 6 is added to the numerator and 14 is added to the denominator, the fraction reduces to $\frac{3}{5}$. The sum of the original numerator and denominator is

A. 60                    D. 90

B. 70                    E. 100

C. 80

**Exercise 9:** Fred leaves Fort Worth by car traveling north. Two hours later, Jim also leaves Fort Worth going north, but 20 mph slower. After six more hours, they are 260 miles apart. Fred's speed is

A. 40 mph               D. 70 mph

B. 50 mph               E. 80 mph

C. 60 mph

**Exercise 10:** The difference in the cost between two books is $8. Together they cost $50. The cost of the less expensive book is

A. $21                   D. $27

B. $23                   E. $29

C. $25

 **Let's look at the answers.**

**Answer 1:** B: Let's break this one down into small pieces. Four more than a number is written as $n + 4$ (or $4 + n$). Seven less than triple the number is $3n - 7$ (the only correct way). *Is* means "equals," so the equation is $n + 4 = 3n - 7$. Solving, we get $n = 5.5$.

**Answer 2:** C: The "setup" to do this problem is $80 \le \dfrac{(98 + 92 + 75 + x)}{4} < 90$. However, $80(4) = 320$ total points for a minimum, and it must be less than $90(4) = 360$ points. So far, Mike has $98 + 92 + 75 = 265$ points; $265 + 66 = 331$ points. (However, answer choices D or E will result in a grade of A, and I'm sure Mike wouldn't object to that.)

**Answer 3:**    **B:** $x - 9 = 9 - x$; $x = 9$.

**Answer 4:**    **C:** If we assume a 100-mile distance, the original trip was 5 hours, and the return trip was 2 hours. $r = \dfrac{d}{t} = \dfrac{200}{7} = 28\dfrac{4}{7}$ mph.

**Answer 5:**    **D:** $3x + 5x + 7x = 180$, so $x = 12°$. The largest angle is $7x = 84°$.

**Answer 6:**    **D:** Dimes are 10 cents each, so the value is $10d$, where $d$ is the number of dimes. Similarly, the value of quarters is $25q$, where $q$ is the number of quarters.

**Answer 7:**    **B:** Let's construct a chart and picture.

|        | r      | t | d        |
|--------|--------|---|----------|
| Slower | x      | 5 | 5x       |
| Faster | x + 30 | 2 | 2(x + 30)|

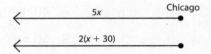

The rate of the slower car is $x$, and the time of the slower car is 7 P.M. −2 P.M., or 5 hours. The rate of the faster car is $x + 30$; and its time is 7 P.M. − 5 P.M., or 2 hours. When they crashed, their distances were equal, so $5x = 2(x + 30)$. Then $x = 20$; and the total distance is $5(20)$ or $2(20 + 30) = 100$ miles.

**Answer 8:**    **A:** The fraction can be written as $\dfrac{2x}{3x}$. So $\dfrac{2x + 6}{3x + 14} = \dfrac{3}{5}$. Cross-multiplying, we get $5(2x + 6) = 3(3x + 14)$, so $x = 12$. The original numerator is $2x = 24$ and the original denominator is $3x = 36$, and their sum is $24 + 36 = 60$.

**Answer 9:**

|      | r      | × | t | = | d         |
|------|--------|---|---|---|-----------|
| Fred | x      |   | 8 |   | 8x        |
| Jim  | x − 20 |   | 6 |   | 6(x − 20) |

**D:** Fred goes at $x$ mph for 8 hours. Jim goes at $x - 20$ mph for 6 hours. Because they are going in the same direction, we get $8x - 6(x - 20) = 260$, or $x = 70$. Notice that 260 is the *difference* in their distances, not the distance they traveled, so we use subtraction.

**Answer 10:**    **A:** Let $x$ = the less expensive book, so $x + 8$ = the more expensive one; $x + (x + 8) = 50$, so $x = 21$. If you were looking for the more expensive book, you could let $x$ = the more expensive book; then $x - 8$ would be the less expensive one!

## MEASUREMENTS

You might want to review some basic measurements and how to convert a few.

Linear: 12 inches = 1 foot; 3 feet = 1 yard; 5,280 feet = 1 mile.

Liquid: 8 ounces = 1 cup; 2 cups = 1 pint; 2 pints = 1 quart; 4 quarts = 1 gallon.

Weight: 16 ounces = 1 pound; 2,000 pounds = 1 ton.

Dry measure: 2 pints = 1 quart; 8 quarts = 1 peck; 4 pecks = 1 bushel. If I love you a bushel and a peck, it would be 5 pecks or 40 dry quarts.

Metric: 1,000 grams in a kilogram; 1,000 milligrams in a gram; 1,000 liters in a kiloliter; 1,000 meters in a kilometer; 1,000 millimeters = 1 meter; 100 centimeters = 1 meter; 10 millimeters = 1 centimeter.

When doing conversions, we pay particular attention to the units, canceling them when doing the multiplication. Sometimes, however, there is confusion about what conversion to use, for example, whether to multiply by $\dfrac{12 \text{ inches}}{1 \text{ foot}}$ or by $\dfrac{1 \text{ foot}}{12 \text{ inches}}$. To avoid this confusion, check to see what units are called for in the answer, and then make sure those units are in the correct place in the conversions.

**Example 14:** Change 30 kilograms 20 grams to milligrams.

**Solution:** The solution is in milligrams, so be sure milligrams is in the numerator (top) of the conversions. If you need further conversions, such as converting kilograms to grams here, be sure the units will cancel with the conversion for milligrams:

$$\frac{30 \text{ kg}}{1} \times \frac{1000 \text{ g}}{1 \text{ kg}} \times \frac{1000 \text{ mg}}{1 \text{ g}} + \frac{20 \text{ g}}{1} \times \frac{1000 \text{ mg}}{1 \text{ g}} = 30{,}000{,}000 + 20{,}000 = 30{,}020{,}000 \text{ mg}$$

**Note**  *Notice that the measurements g and kg cancel.*

**Example 15:**  Change 90 miles per hour into feet per second.

**Solution:**    The answer is in feet per second, so feet should be in the numerator (top) of any conversion, and seconds should be in the denominator (bottom) of any conversion:

$$\frac{90 \text{ miles}}{\text{hour}} \times \frac{1 \text{ hour}}{60 \text{ minutes}} \times \frac{1 \text{ minute}}{60 \text{ seconds}} \times \frac{5280 \text{ feet}}{1 \text{ mile}} = \frac{132 \text{ feet}}{\text{sec}}$$

 *Each fraction after the first is equivalent to 1. When we multiply by 1, the value doesn't change. Again, the measurements cancel, and we wind up with feet per second.*

## INTEREST

Last, we need to talk a little about simple interest. We know that the interest is equal to the principal times (annual) rate times time (in years). In symbols $i = prt$. The total amount of money is the principal plus the interest, or $A = p + i = p + prt = p(1 + rt)$.

**Example 16:**  Suppose we invest $20,000 at simple interest at 12% for 3 months. How much money do we have?

**Solution:**    $A = p + prt = 20,000 + 20,000(.12)\left(\frac{1}{4}\right) = \$20,600$. Of course, we wouldn't normally invest at simple interest. However, a safe 12% interest would be great.

## Chapter 7 Quiz

For questions 1–12: Write in symbols: Do *not* solve any problems.

1. 2x less than y.

2. The product of m and the sum of m and n.

3. The difference of b and c divided by y.

4. The product of v and m from the difference between r and s.

5. Five less than twice a number is the same as 9 less four times the same number.

6. The volume V of a cone is equal to the product of one-third pi, π, the radius r squared, and the height h.

7. The age $A$ of an adult is at least 21.

8. The height $h$ of people in the class is between 66 inches and 80 inches.

9. The area of a trapezoid $A$ equals one-half the product of the height $h$ and the sum of the bases $b_1$ and $b_2$.

10. The amount of money $A$ equals the sum of the principal $p$ and the product of the principal, the rate $r$, and the time $t$.

11. Let $x$ = the larger number and $y$ = the smaller number. Write the expression that says the sum of twice the larger number and triple the smaller number is the same as the product of the two numbers.

12. Let $x$ = the larger number and $y$ = the smaller number. Write the expression that says the quotient of the larger number divided by the smaller number is at most 12.

13. Change 4 tons 5 pounds 3 ounces to ounces.

14. Change 543 meters to centimeters and kilometers.

15. A board is divided into the ratio of 4:7:9. If the board is 160 inches long, find the length of each piece.

16. You drove 3 hours at 40 miles per hour and 6 hours at 60 mph. Find your average speed.

17. From problem 10, if you invest $2,000 for 10 years at 8% interest, find your total after 10 years.

18. One number is 10 more than another. The total is 84. Find the numbers.

19. The cost of an item is $6. The cost of another item is $8. If I buy one at $6, what is the largest number of $8 items I can buy if I have $100?

20. You want an average between 90 and 95 inclusive. You received 88, 95, and 99 on three tests. What grades can you get on the fourth test so your grade will be between 90 and 95?

## Answers to Chapter 7 Quiz

1. $y - 2x$.

2. $m(m + n)$.

3. $\dfrac{b - c}{y}$.

4. $(r - s) - vm$.

5. $2n - 5 = 9 - 4n$.

6. $V = \dfrac{1}{3}\pi r^2 h$.

7. $A \geq 21$.

8. $66 < h < 80$.

9. $A = \dfrac{1}{2}h(b_1 + b_2)$.

10. $A = p + prt$.

11. $2x + 3y = xy$.

12. $\dfrac{x}{y} \le 12$.

13. $4(2000)(16) + 5(16) + 3 = 128{,}083$.

14. a. $543 \times 100 = 54{,}300$ cm; b. $\dfrac{543}{1000} = .543$ km.

15. $4x + 7x + 9x = 20x = 160$; $x = 8$; so the pieces in inches are $4(8) = 32$, $7(8) = 56$, and $9(8) = 72$.

16. $\dfrac{3(40) + 6(60)}{9} = \dfrac{480}{9} = \dfrac{160}{3} = 53\dfrac{1}{3}$ mph.

17. $A = 2{,}000 + 2{,}000(.08)(10) = \$3{,}600$.

18. $x + (x + 10) = 84$; $x = 37$ and $x + 10 = 47$.

19. $6 + 8x \le 100$; solving we get $x \le 11\dfrac{3}{4}$. The answer is 11 because you can't buy three-fourths of an item.

20. $90 \le \dfrac{88 + 95 + 99 + x}{4} \le 95$; so $360 \le 282 + x \le 380$; and $78 \le x \le 98$.

**Answer to "Bob Asks":** Bookkeeper

# CHAPTER 8: *Two or More Unknowns*

"*Can you name two English words with UU in them?*"

## SOLVING SIMULTANEOUS EQUATIONS

There are three ways to solve two equations in two unknowns that you may have seen before and two that you probably haven't. There is solving by graphing. We will do that method in the next chapter. The second is by substitution, which we will go over perhaps a little more than you've previously seen. The third is the elimination method, sometimes called the adding and subtracting method. The two that you probably have not seen involve matrices and determinants. The one by matrices, fortunately, you do not need. The second, by determinants, is related to matrices and a great way to solve two equations in two unknowns. We shall see this in a later chapter.

### Substitution

In substitution, we find an unknown with a coefficient of 1, solve for that variable, and substitute it in the other equation.

**Example 1:** Solve for $x$ and $y$:

$$3x + 4y = 4 \qquad (1)$$
$$x - 5y = 14 \qquad (2)$$

**Solution:** In equation (2) $x = 5y + 14$. Substituting this into equation (1), we get $3(5y + 14) + 4y = 4$. Solving, we get $y = -2$; then $x = 5y + 14 = 5(-2) + 14 = 4$. The answer is $x = 4$ and $y = -2$.

We can check by substituting these values into the original equations to see that they are solutions to both equations.

**Note** *If the coefficient of none of the terms is 1, the elimination method is better.*

## Elimination

There are several ways to eliminate one of the variables, as seen in the following examples. Once we have eliminated one of the variables, we can solve for the other variable by using substitution.

**Example 2:**    $2x + 3y = 12$

$5x - 3y = 9$

**Solution:**    If we add the equations, term by term, we can eliminate the $y$ term. We get $7x = 21$, so $x = 3$; substituting $x = 3$ into either equation, we get $y = 2$. So the answer is $x = 3, y = 2$.

If addition doesn't work, try subtraction.

**Example 3:**    $5x + 4y = 14$

$5x - 2y = 8$

**Solution:**    By subtracting, we get $6y = 6$, so $y = 1$; then by substituting, we get $x = 2$.

If adding or subtracting doesn't work, we must find two numbers that, when we multiply the first equation by one of them and the second equation by the other, and then add (or subtract) the resulting equations, one letter is eliminated.

**Example 4:**    $5x + 3y = 11$          (1)

$4x - 2y = 22$          (2)

**Solution:**    To eliminate $x$, multiply equation (1) by 4 and equation (2) by $-5$; then add.

$4(5x + 3y) = 4(11)$       or       $20x + 12y = 44$

$-5(4x - 2y) = -5(22)$       or       $-20x + 10y = -110$

Adding, we get $22y = -66$; so $y = -3$. We could substitute now, but we could also eliminate $y$ by multiplying the original equation (1) by 2 and equation (2) by 3. Let's do that.

$2(5x + 3y) = 2(11)$       or       $10x + 6y = 22$

$3(4x - 2y) = 3(22)$       or       $12x - 6y = 66$

Adding, we get $22x = 88$, so $x = 4$. The answer is $x = 4, y = -3$.

**Example 5:**     Solve the pairs of equations:

<u>Problem</u>                                     <u>Solution</u>

a.  $x - y = 20$                          $x = 15, y = -5$ $(15, -5)$

  $x + y = 10$

b.  $5x - y = 23$                        $(4, -3)$

  $x - y = 7$

c.  $4x + 3y = 24$                       $(3, 4)$

  $2x - 5y = -14$

d.  $-3x + 5y = 4$                       $(-3, -1)$

  $-2x + y = 5$

**Q  Let's do some multiple-choice exercises.**

**Exercise 1:**     If $x = y + 3$ and $y = z + 7$, $x$ (in terms of $z$) =

A. $z - 10$                      D. $z + 4$

B. $z - 4$                       E. $z + 10$

C. $z$

**Exercise 2:**     Two apples and 3 pears cost 65 cents, and 5 apples and 4 pears cost $1.10. Find the cost, in cents, of one pear:

A. 10                            D. 25

B. 15                            E. 30

C. 20

**Exercise 3:**     As in Exercise 2, 2 apples and 3 pears cost 65 cents, and 5 apples and 4 pears cost $1.10. Find the cost, in cents, of one pear and one apple together:

A. 10                            D. 25

B. 15                            E. 30

C. 20

**Exercise 4:**    Find $x + y$:

$7x + 4y = 27$

$x - 2y = -3$

A. 1                                          D. 7

B. 3                                          E. 9

C. 5

**Exercise 5:**    For lunch, Ed buys 3 hamburgers and 1 soda for $12.50, and Mei buys 1 hamburger and 1 soda for $5.60. How much does Ed pay for his hamburgers?

A. $2.15                                  D. $10.35

B. $3.45                                  E. $18.10

C. $6.90

**Exercise 6:**    The product of 4 and the sum of $x$ and $y$ is at least as large as the quotient of $a$ divided by $b$. This can be written as

A. $4x + y - \dfrac{a}{b} \geq 0$          D. $\dfrac{a}{b} - 4x + 4y < 0$

B. $x + 4y - \dfrac{a}{b} > 0$          E. $4(x + y) + \dfrac{a}{b} \geq 0$

C. $4(x + y) - \dfrac{a}{b} \geq 0$

Ⓐ   **Let's look at the answers.**

**Answer 1:**    E: $x = y + 3 = (z + 7) + 3 = z + 10$.

**Answer 2:**    B:

$2a + 3p = 65$

$5a + 4p = 110$

In solving for $p$, eliminate $a$ by multiplying the top equation by 5 and the bottom by $-2$.

$5(2a + 3p) = 5(65)$          or          $10a + 15p = \phantom{-}325$

$-2(5a + 4p) = -2(110)$      or          $-10a - \phantom{0}8p = -220$

Adding, we get $7p = 105; p = 15$.

**Answer 3:**    **D:** Much more often, we get a problem like this. The equations are the same as in Exercise 2, but rather than asking for the cost of one apple or the cost of one pear, this exercise asks for the cost of one apple plus one pear. The trick is simply to add the original equations. We then get $7a + 7p = 175$. Dividing both sides by 7, we get $a + p = 25$.

**Answer 4:**    **C:** Less frequently, when adding doesn't work, try subtracting. If we subtract, the difference becomes $6x + 6y = 30$. So $x + y = 5$.

**Answer 5:**    **D:** The equations are

$$3h + s = 12.50$$

$$h + s = 5.60$$

Subtracting, we get $2h = 6.90$, so $h = 3.45$. Ed's three hamburgers cost $10.35.

**Answer 6:**    **C:** $4(x + y) \geq \dfrac{a}{b}$. After rearranging, only answer choice C is correct.

# Chapter 8 Quiz

For questions 1–14, solve for $x$ and $y$:

1.  $x - y = 12$

    $x + y = 2$

2.  $4x - 3y = 25$

    $x - 3y = 10$

3.  $2x - 5y = -6$

    $5x - 7y = -15$

4.  $4x + 6y = 3$

    $8x - 9y = -1$

5.  $\dfrac{x}{4} - \dfrac{y}{3} = 2$

    $\dfrac{x}{6} - \dfrac{5y}{3} = -3$

6.  $y = 3x + 2$

    $4x + 3y = 32$

7.  $ax + by = f$

    $cx + dy = g$

8.  $4x - 6y = 12$

    $6x - 9y = 18$

9.  $2x - 7y = 3$

    $-6x + 21y = 5$

10. Sue bought $x$ dresses at $70 dollars and $y$ shoes at $50 a pair. She has at most $1,000 to spend. Write this as an inequality.

11. The amount of $x$ $10 bills, $y$ $50 bills, and $z$ $100 bills totals more than $10,000. Write the inequality.

12. The cost of 4 hamburgers and 6 hot dogs is twenty-nine dollars. The cost of 7 hamburgers and 3 hot dogs is thirty-two dollars. Find the cost of one hamburger and one hot dog.

13. I have 30 dimes and nickels that total $2.40. How many of each coin do I have?

14. Four times of sum of $x$ and $y$ minus three times the difference of $x$ and $y$ is 19. The sum of the numbers is 7. Find the numbers.

15. At a modern brand new major league ballpark, believe it or not, the cost of 12 deluxe box seats and 5 "ordinary" box seats cost $11,450 (for one game!!!!), and the cost of 7 deluxe box seats and 3 "ordinary" box seats cost $6,700. Find the cost of one deluxe and one ordinary box seat.

## Answers to Chapter 8 Quiz

1.  Add them; $2x = 14$; $x = 7$; then substitute: $(7, -5)$.

2.  Subtract; $3x = 15$; $x = 5$; substitute $\left(5, -\dfrac{5}{3}\right)$.

3.  Multiply the top equation by 5 and the bottom equation by $-2$; $-11y = 0$; $y = 0$; $(-3, 0)$.

4.  Multiply the top equation by $-2$ and add; $-21y = -7$; $y = \dfrac{1}{3}$; $\left(\dfrac{1}{4}, \dfrac{1}{3}\right)$.

5.  First, multiply the top equation by 12 and the bottom equation by 6 to get rid of the fractions; we then have $3x - 4y = 24$ and $x - 10y = -18$. Multiply the second equation by $-3$ and add; $26y = 78$; $y = 3$; $(12, 3)$.

6.  Use substitution and add; $4x + 3(3x + 2) = 32$; $13x + 6 = 32$; $x = 2$; $(2, 8)$.

7.  For $y$, multiply the top equation by $c$ and the bottom by $-a$; $y = \dfrac{ag - cf}{ad - bc}$; for $x$, multiply the top equation by $d$ and the bottom by $-b$ and add; $x = \dfrac{df - bg}{ad - bc}$.

8.  Multiply the top equation by 3 and the bottom equation by $-2$; after you add, you will get $0 = 0$, which means there isn't just one answer. Both equations are the same, so any answer to one is an answer to the other. You may want to look at this one again after the next chapter (you can see the answer by graphing carefully).

9.  If you multiply the top equation by 3 and then add the equations, you get $0 = 14$, which is never true. This means the two equations have no common points. If you graph the equations, the lines are parallel.

10. $70x + 50y \le \$1,000$.

11. $10x + 50y + 100z > 10,000$.

12. $4H + 6D = 29$ and $7H + 3D$ (this problem has depth) $= 32$; multiply the bottom equation by $-2$ and add them; $-10H = -35$; $H = \$3.50$; $D = \$2.50$. So the cost is $\$6.00$.

13. $N + D = 30$; $5N + 10D = 240$; $5(30 - D) + 10D = 240$; $150 - 5D + 10D = 240$; $5D = 90$; $D = 18$; $N = 12$. Check? $18(10) + 12(5) = 240$ pennies.

14. $4(x + y) - 3(x - y) = x + 7y = 19$ and $x + y = 7$; so $7 - y + 7y = 19$; $6y = 12$; $y = 2$; $x = 5$

15. Let $D =$ deluxe and $O =$ ordinary (OK, you shouldn't use $O$, but I will); $12D + 5O = 11,450$, and $7D + 3O = 6,700$; multiply the first equation by 3 and the second by $-5$; $36D + 15O = 34,350$, and $-35D - 15O = -33,500$; $D = \$850.00$! and $O = \$250$! Take me out to the ball game! After this question, I'm ready for anything!

Let's take a break from algebra, and take a look at points, lines, and shapes.

**Answer to "Bob Asks":** Vacuum and continuum

*"A clock rings the number of hours each hour and it rings once on the half-hour. At six o'clock the bell rings six times; at 6:30 once, and at seven o'clock the bell rings seven times. John comes home to hear the bell ring once. A half-hour later it rings once, a half-hour later it rings once again, and a half-hour later, it rings once again. If the clock is not broken, when did John come home?"*

**This** topic used to be part of a course called analytic geometry (algebraic geometry). We'll start at the beginning.

## POINTS IN THE PLANE

We start with a **plane**—a two-dimensional space, like a piece of paper. On this plane, we draw two perpendicular lines, or **axes**. The x-axis is horizontal; the y-axis is vertical. Positive x is to the right; negative x is to the left. Positive y is up; negative y is down. Points in the plane are indicated by **ordered pairs** (x, y). The x number, called the **first coordinate** or **abscissa**, is always given first; the y number, called the **second coordinate** or **ordinate**, is always given second. Here are some points on the plane.

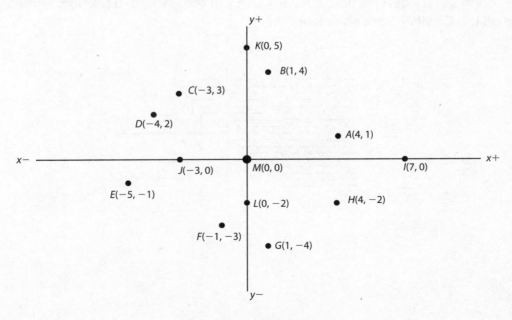

Note the following:

- For any point on the *x*-axis, the *y*-coordinate always is 0.

- For any point on the *y*-axis, the *x*-coordinate is 0.

- The point where the two axes meet, (0, 0), is called the **origin**.

- The axes divide the plane into four **quadrants**, usually written with roman numerals, starting in the upper right quadrant and going counterclockwise.

  - In quadrant I, $x > 0$ and $y > 0$.

  - In quadrant II, $x < 0$ and $y > 0$.

  - In quadrant III, $x < 0$ and $y < 0$.

  - In quadrant IV, $x > 0$ and $y < 0$.

In the following figure, we have drawn the line $y = x$. For every point on this line, the first coordinate has the same value as the second coordinate, or $y = x$. If we shade the area above this line, $y > x$ in the shaded portion. Similarly, $x > y$ in the unshaded portion. Sometimes, questions on the COMPASS ask about this.

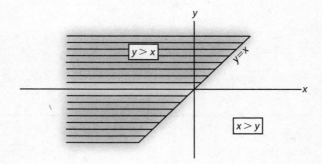

The following figure shows symmetry of the point $(a, b)$ about the $x$-axis, $y$-axis, and the origin. Suppose $(a, b)$ is in quadrant I. Then $(-a, b)$ would be in quadrant II, $(-a, -b)$ would be in quadrant III, and $(a, -b)$ would be in quadrant IV, as pictured.

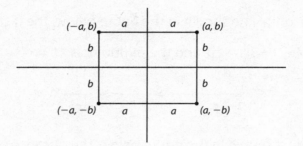

# LINES

## Distance and Midpoint

The formulas for distance and midpoint look a little complicated, but they are fairly easy to use. It just takes practice.

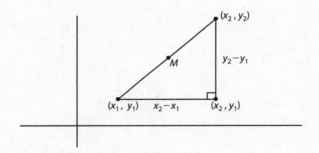

To find the **distance** between two points $(x_1, y_1)$ and $(x_2, y_2)$ on a plane, we must use the distance formula:

$$d = \sqrt{(x_2 - x_1)^2 + (y_2 - y_1)^2}$$

The distance formula is just the Pythagorean Theorem (discussed in the next chapter).

Distances are always positive. You may be six feet tall, but you cannot be minus six feet tall.

The **midpoint** of a line between two points $(x_1, y_1)$ and $(x_2, y_2)$ on a plane is given by the coordinates

$$M = \left( \frac{x_1 + x_2}{2}, \frac{y_1 + y_2}{2} \right)$$

In one dimension, if the line is horizontal, these formulas simplify to $d = |x_2 - x_1|$ and $M = \dfrac{x_1 + x_2}{2}$.

Similarly, if the line is vertical, these formulas simplify to $d = |y_2 - y_1|$ and $M = \dfrac{y_1 + y_2}{2}$.

For example, for the horizontal line shown in the figure below, the distance between the points is $d = |x_2 - x_1| = |7 - (-3)| = 10$, and the midpoint is $M = \dfrac{x_1 + x_2}{2} = \dfrac{-3 + 7}{2} = 2$.

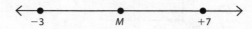

Similarly, for the vertical line shown in the figure below, the distance between the points is $d = |y_2 - y_1| = |-4 - (-8)| = 4$, and the midpoint is $M = \dfrac{y_1 + y_2}{2} = \dfrac{(-8) + (-4)}{2} = -6$.

## Slope

The **slope** of a line tells by how much the line is "tilted" compared to the $x$-axis. The formula for the slope of a line is

$$m = \frac{\text{rise}}{\text{run}} = \frac{\text{change in } y}{\text{change in } x} = \frac{y_2 - y_1}{x_2 - x_1},$$

where $(x_1, y_1)$ and $(x_2, y_2)$ are any two points on the line.

Note the following facts about the slope of a line, as shown in the figure below:

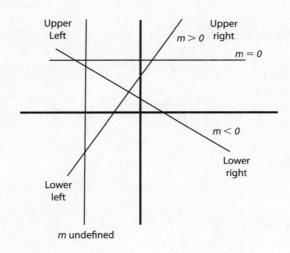

- The slope is positive if the line goes from the lower left to the upper right.
- The slope is negative if it goes from the upper left to the lower right.
- Horizontal lines have zero slope.
- Vertical lines have no slope or undefined slope or "infinite" slope.

**Example 1:**  Find the distance, slope, and midpoint for the line segment joining these points:

**a.** (2, 3) and (6, 8)      **c.** (7, 3) and (4, 3)

**b.** (4, −3) and (−2, 0)      **d.** (2, 1) and (2, 5)

**Solutions:**

**a.** We let $(x_1, y_1) = (2, 3)$ and $(x_2, y_2) = (6, 8)$, although the other way around is also okay.

Then

$$\text{Distance} = d = \sqrt{(x_2 - x_1)^2 + (y_2 - y_1)^2} = \sqrt{(6-2)^2 + (8-3)^2} = \sqrt{41}$$

$$\text{Slope} = m = \frac{y_2 - y_1}{x_2 - x_1} = \frac{8-3}{6-2} = \frac{5}{4}$$

$$\text{Midpoint} = M = \left(\frac{x_1 + x_2}{2}, \frac{y_1 + y_2}{2}\right) = \left(\frac{2+6}{2}, \frac{3+8}{2}\right) = (4, 5.5)$$

Notice that the slope is positive; the line segment goes from the lower left to the upper right.

**b.** We let $(x_1, y_1) = (4, -3) =$ and $(x_2, y_2) = (-2, 0)$.

$$\text{Distance} = d = \sqrt{(-2-4)^2 + (0-(-3))^2} = \sqrt{45}$$

$$= \sqrt{3 \times 3 \times 5} = 3\sqrt{5}$$

$$\text{Slope} = m = \frac{0 - (-3)}{-2 - 4} = -\frac{1}{2}$$

$$\text{Midpoint} = M = \left(\frac{4 + (-2)}{2}, \frac{-3 + 0}{2}\right) = (1, -1.5)$$

Notice that the slope is negative; the line segment goes from the upper left to the lower right.

**c.** We let $(x_1, y_1) = (7, 3) =$ and $(x_2, y_2) = (4, 3)$.

It is a one-dimensional horizontal distance (the $y$ values are the same), so $d = |4 - 7| = 3$

$$\text{Slope} = m = \frac{3 - 3}{4 - 7} = \frac{0}{-3} = 0$$

$$\text{Midpoint} = M = \left(\frac{7 + 4}{2}, \frac{3 + 3}{2}\right) = (5.5, 3)$$

Notice that the horizontal line segment has slope $m = 0$.

**d.** We let $(x_1, y_1) = (2, 1) =$ and $(x_2, y_2) = (2, 5)$.

Again, this is a one-dimensional distance (but vertical since the $x$ values are the same), so $d = |5 - 1| = 4$

$$\text{Slope} = m = \frac{5 - 1}{2 - 2} = \frac{4}{0}, \text{ undefined (cannot divide by 0)}$$

$$\text{Midpoint} = M = \left(\frac{2 + 2}{2}, \frac{5 + 1}{2}\right) = (2, 3)$$

Notice that the slope of the vertical line segment is undefined.

**Now let's do some multiple-choice exercises.**

**Exercise 1:** The coordinates of *P* are (*j*, *k*). If s < *k* < *j* < *r*, which of the points shown in the figure could have the coordinates (*r*, *s*)?

**A.** A

**D.** D

**B.** B

**E.** E

**C.** C

Use the figure below for Exercises 2 and 3.

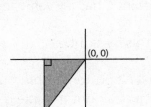

**Exercise 2:** Which of the following points is inside the triangle?

**A.** (−3, 6)

**D.** (−3, −4)

**B.** (−5, −5)

**E.** (−1, −3)

**C.** (−2, −5)

**Exercise 3:** The area of the triangle is

**A.** 6

**D.** 24

**B.** 12

**E.** 48

**C.** 18

**Exercise 4:** *M* is the midpoint of the horizontal line segment *AB*. If the coordinates of *A* are (*m*, −*n*), then the coordinates of *B* are

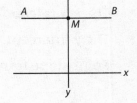

**A.** (*m*, *n*)

**D.** (*n*, *m*)

**B.** (−*m*, *n*)

**E.** (−*n*, −*m*)

**C.** (−*m*, −*n*)

**Exercise 5:**    In the given figure, *AB* is parallel to (∥) the *x*-axis and *PQ* = *AB*. The coordinates of point *A* are

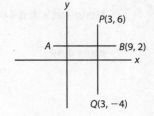

A. (−1, 2)          D. (9, 0)

B. (1, 2)           E. (−9, 2)

C. (9, −8)

    **Let's look at the answers.**

**Answer 1:**    **C:** For points *C*, *D*, and *E*, the *x* value is bigger than the *x* value of *P*; of these, only point *C* has a *y* value less than the *y* value of *P*.

**Answer 2:**    **D:** You can do this by sight.

**Answer 3:**    **B:** We really haven't gotten to this, but I asked it because we have the picture. The area of a right triangle is half the horizontal distance times the vertical distance.

$$A = \frac{1}{2}bh = \frac{1}{2} \times 4 \times 6 = 12$$

**Answer 4:**    **C:** Slightly tricky. Point *B* has the same *y* value as *A*, but its *x* value is the negative of the *x* value for *A*. Note that the actual values for *m* and *n* must be negative numbers in this figure.

**Answer 5:**    **A:** The length of *PQ* = 10. For the length of *AB* to be 10, *A* must be (−1, 2) since $|9 − (−1)| = 10$.

## Standard Equation of a Line

Let's go over the facts we need.

- **Standard form** of the line: *Ax* + *By* = *C*; *A*, *B* both ≠ 0.

- The ***x*-intercept**, the point at which the line hits the *x*-axis, occurs when *y* = 0.

- The ***y*-intercept**, the point at which the line hits the *y*-axis, occurs when *x* = 0.

- **Point-slope form** of a line: Given slope *m* and point $(x_1, y_1)$, the point-slope form of a line is $m = \dfrac{y - y_1}{x - x_1}$.

- **Slope-intercept form** of a line: $y = mx + b$, where $m$ is the slope and $(0, b)$ is the $y$-intercept.

- Lines of the form:

  $y =$ constant, such as $y = 2$, are lines parallel to the $x$-axis; the equation of the $x$-axis is $y = 0$.

  $x =$ constant, such as $x = -3$, are lines parallel to the $y$-axis; the equation of the $y$-axis is $x = 0$.

  $y = mx$ are lines that pass through the origin.

**Example 2:** For $Ax + By = C$, find the $x$ and $y$ intercepts.

**Solution:** The $y$-intercept means $x = 0$; so $y = \dfrac{C}{B}$, and the $y$-intercept is $\left(0, \dfrac{C}{B}\right)$.

The $x$-intercept means $y = 0$; so $x = \dfrac{C}{A}$, and the $x$-intercept is $\left(\dfrac{C}{A}, 0\right)$.

**Example 3:** For $3x - 4y = 7$, find the $x$ and $y$ intercepts.

**Solution:** For the $y$-intercept, $x = 0$; so $y = \dfrac{7}{-4}$, and the $y$-intercept is $\left(0, -\dfrac{7}{4}\right)$.

For the $x$-intercept, $y = 0$; so $x = \dfrac{7}{3}$, and the $x$-intercept is $\left(\dfrac{7}{3}, 0\right)$.

**Example 4:** Given a line with $m = \dfrac{3}{2}$ and containing point $(5, -7)$, write the equation of the line in standard form.

**Solution:** $m = \dfrac{y - y_1}{x - x_1}$, so $\dfrac{3}{2} = \dfrac{y - (-7)}{x - 5}$. Cross-multiplying, we get $3(x - 5) = 2(y + 7)$, or $3x - 2y = 29$.

**Example 5:** Given points $(3, 6)$ and $(7, 11)$ on a line, write the equation of the line in slope-intercept form.

**Solution:** $y = mx + b$. $m = \dfrac{11 - 6}{7 - 3} = \dfrac{5}{4}$, and we will use point $(3, 6)$, so $x = 3$ and $y = 6$. Therefore, $6 = \dfrac{5}{4}(3) + b$, and $b = \dfrac{9}{4}$. So the line is $y = \dfrac{5}{4}x + \dfrac{9}{4}$.

**Example 6:** Sketch lines $x = -3$, $y = 8$, and $y = \frac{2}{3}x$.

**Solution:**

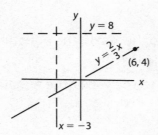

**Q** **Let's do a couple more multiple-choice exercises.**

**Exercise 6:** A line with the same slope as the line $y = \frac{2}{3}x - 5$ is

    **A.** $2x = 18 - 3y$         **D.** $-2x - 3y = 14$

    **B.** $2x + 3y = 6$           **E.** $2y = 12 - 3x$

    **C.** $2x - 3y = 6$

**Exercise 7:** Find the area of the triangle formed with the positive $x$-axis, positive $y$-axis, and the line through the point (3, 4) with slope $-2$. The area is

    **A.** 5                **D.** 50

    **B.** 15              **E.** 10

    **C.** 25

**A** **Let's look at the answers.**

**Answer 6:** C: We have to solve for $y$ in each case, but we are interested in only the coefficient of $x$. The only answer choice that works is C.

**Answer 7:** C: You must visualize the figure.

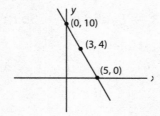

Using the point-slope form, the equation of the line is $-2 = \dfrac{y - 4}{x - 3}$.

If we let $x = 0$, the $y$-intercept is 10. If we let $y = 0$, the $x$-intercept is 5. The area of the triangle is one-half the $x$ distance times the $y$ distance.

$$\text{Area} = \frac{1}{2}ab = \frac{1}{2} \times 5 \times 10 = 25.$$

## Parallel and Perpendicular Lines

We say two lines are **parallel** if they have the **same slope**.

 *There is a special case that two lines are parallel also if they are parallel to the y-axis. We need this case since lines parallel to the y-axis are said to have infinite slope. Since infinity is not a number, we can't say infinity equals infinity. In fact, as a teaser for those who care, there is more than one infinity. In fact, there are an infinite number of infinities. Let's move on!*

We say two lines are **perpendicular** if their slopes are **negative reciprocals** of each other. Again, there is a special case for this. Two lines are also perpendicular if one is parallel to the $x$-axis (slope 0) and one is parallel to the $y$-axis (infinite slope).

**Example 7:**     Given line $L$: $7x + 8y = 9$.

**a.** Find the equation in standard form of a line parallel to $L$ through the point (2, 3).

**b.** Find the equation in standard form of a line perpendicular to $L$ through the point (5, −6).

**Solution:**     In each case we must find the slope. Solving for $y$, we get $y = -\dfrac{7}{8}x + \dfrac{9}{8}$.

**a.** Since the slope is the coefficient of $x$, $-\dfrac{7}{8}$, the parallel line will have the same slope, $-\dfrac{7}{8}$, and the point (2, 3). We can use the point-slope form: $m = \dfrac{y - y_1}{x - x_1}$, or $\dfrac{-7}{8} = \dfrac{y - 3}{x - 2}$. Cross-multiplying, we get $-7(x - 2) = 8(y - 3)$; $-7x + 14 = 8y - 24$; rearranging we get $7x + 8y = 38$.

**b.** The perpendicular line will have a slope that is the negative reciprocal of $\dfrac{-7}{8}$, or $+\dfrac{8}{7}$. The point-slope form is $m = \dfrac{y - y_1}{x - x_1}$, or $\dfrac{8}{7} = \dfrac{y - (-6)}{x - 5}$. Then $8(x - 5) = 7(y + 6)$, or $8x - 7y = 82$.

## Solving Two Equations in Two Unknowns by Graphing

It may be necessary to solve two equations in two unknowns by graphing. What you must do is graph each line. The point where they meet is the solution to the problem.

**Example 8:** Solve by graphing: $2x + y = 8$
$$2x + 3y = 12$$

**Solution:** We graph by using intercepts.

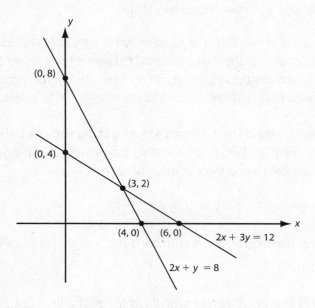

$2x + y = 8$: If $x = 0$, then $y = 8$, so the $y$-intercept is $(0, 8)$; if $y = 0$, then $x = 4$, so the $x$-intercept is $(4, 0)$. Graph the line through these two points.

$2x + 3y = 12$: If $x = 0$, then $y = 4$, and the $y$-intercept is $(0, 4)$; if $y = 0$, then $x = 6$, and the $x$-intercept is $(6, 0)$. Graph the line through these two points. The lines meet at the point $P(3, 2)$. You can check $x = 3$ and $y = 2$ in both equations. On the COMPASS, this method will rarely be used since drawing pictures would lead to inaccuracies and take more time.

## EQUATION OF A CIRCLE

Here we do an algebraic version of the circle. We will do more, like in your geometry class, in Chapter 12.

A circle is the set of all points $(x, y)$ at a given distance $r$ (called the radius) from a given point $(h, k)$ called the center.

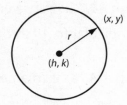

Finding the radius means using the distance formula, so we get the following:

$$r = \sqrt{(x - h)^2 + (y - k)^2}$$

**Note**  $(h, k)$ is the center, a fixed point.

**Note**  $(x, y)$ is any point on the circumference.

No one uses this form of the circle. Instead, we square both sides. So the equation of the circle is $(x - h)^2 + (y - k)^2 = r^2$.

**Example 9:** What is the center and radius of the circle $(x - 5)^2 + (y + 6)^2 = 11$.

**Solution:** Using the equation of a circle, we see that $h = 5$, $k = -6$, and $r^2 = 11$. So the center is the point $(5, -6)$ and the radius is $\sqrt{11}$.

**Example 10:** Write the equation of a circle with radius 7 and center $(-8, 9)$.

**Solution:** Substituting directly into the equation of a circle, we get $(x + 8)^2 + (y - 9)^2 = 49$.

**Note**  *If you get any questions on these last two topics, especially circles, sketch the picture. You'll be surprised how many times the picture will give you the answer.*

## Chapter 9 Quiz

For questions 1–6, let point $A$ be $(4, -7)$ and point $B$ be $(-4, -11)$.

1. Find the midpoint of $AB$.
2. Find the slope of $AB$.
3. Find the length of $AB$.
4. Find the equation of the line through $AB$. (Give answer in standard form.)
5. Find the equation of the line perpendicular to $AB$ through $(-2, 1)$. (Give the answer in point-slope form.)
6. Find the equation of the circle that has $AB$ as its diameter.
7. What is the equation of any line perpendicular to the $y$-axis?
8. The line $3x - 4y = 5$ goes through which quadrants?
9. Write the equation of the line through $(5, 7)$ parallel to the $y$-axis.
10. Write the equation of the line with $x$-intercept 4 and $y$-intercept 2.
11. Write the equation of a circle with center $(4, -6)$ and radius 3.
12. Find the center and radius of a circle if the equation is $(x - 11)^2 + (y + 4)^2 = 13$.
13. Write the equation of a circle with center at $(-5, -4)$, tangent to (just touching) the $y$-axis.

For questions 14 and 15, $L$ is $4y - 8x = 27$.

14. Find the equation of a line parallel to $L$ through $(-8, 3)$. (Give the answer in point-slope form.)
15. Find the equation of the line perpendicular to $L$ though $(-5, -7)$. (Give the answer in point-slope form.)

For questions 16–20, let point $M = (a, b)$ and point $N = (5a, 7b)$.

16. Find the midpoint of $MN$.
17. Find the slope of $MN$.
18. Find the distance between $M$ and $N$.
19. Find the equation of the line parallel to $MN$ through the point $(c, -d)$. (Give the answer in point-slope form.)
20. Find the equation of the circle with $MN$ as the diameter.

## Answers to Chapter 9 Quiz

1. $\left( \dfrac{4 + (-4)}{2}, \dfrac{(-7) + (-11)}{2} \right) = (0, -9)$.

2. $m = \dfrac{y_2 - y_1}{x_2 - x_1} = \dfrac{-11 - (-7)}{-4 - 4} = \dfrac{-4}{-8} = \dfrac{1}{2}$.

3. $\sqrt{(x_2 - x_1)^2 + (y_2 - y_1)^2} = \sqrt{(-4 - 4)^2 + [-11 - (-7)]^2} = \sqrt{(-8)^2 + (-4)^2} = \sqrt{80} = 4\sqrt{5}$.

4. From question #2, the slope is $\frac{1}{2}$. Then $\frac{1}{2} = \frac{y - (-7)}{x - 4}$, or $x - 2y = 18$.

5. For a perpendicular line, the slope is the negative reciprocal, so $-2 = \frac{y - 1}{x - (-2)} = \frac{y - 1}{x + 2}$.

   On a class test, you must leave it in the form the teacher asks. On the COMPASS, you must match the form of your answer to the test answer choices. Don't do any more work than you have to. It's the way of the mathematician.

6. $r = \frac{1}{2}d = \frac{1}{2}(4\sqrt{5}) = 2\sqrt{5}$. The midpoint is the center (from question #1), $(0, -9)$. The equation of the circle is therefore $x^2 + (y + 9)^2 = 20$.

7. $y =$ any number.

8. I, III, and IV (sketch the picture by using intercepts).

9. $x = 5$.

10. The points are $(4, 0)$ and $(0, 2)$; $m = \frac{2 - 0}{0 - 4} = -\frac{1}{2}$. In slope-intercept form, $y = -\frac{1}{2}x + 2$.

11. $(x - 4)^2 + (y + 6)^2 = 3^2$

12. $(11, -4), r = \sqrt{13}$.

13. $(x + 5)^2 + (y + 4)^2 = (4)^2$.

14. $m = 2; 2 = \frac{y - 3}{x + 8}$.

15. The slope of $L$ is 2; the slope of a line perpendicular is $-\frac{1}{2}$. The equation is $-\frac{1}{2} = \frac{y + 7}{x + 5}$.

16. $(3a, 4b)$.

17. $\frac{6b}{4a} = \frac{3b}{2a}$.

18. $\sqrt{(5a - a)^2 + (7b - b)^2} = \sqrt{16a^2 + 36b^2} = 2\sqrt{4a^2 + 9b^2}$.

19. From question 17, the slope of $MN$ is $\frac{3b}{2a} = \frac{y + d}{x - c}$. Letters are nice—no arithmetic to do.

20. From question 18, we have $d = 2\sqrt{4a^2 + 9b^2}$, and for the radius, we divide by 2. From question 16, the center is $(3a, 4b)$. So the equation is $(x - 3a)^2 + (y - 4b)^2 = 4a^2 + 9b^2$.

Let's finally get to angles and triangles.

**Answer to "Bob Asks":** John comes in just in time to hear the last ring of 12:00. This question shows you that you must understand the question before you answer it.

# CHAPTER 10: *All About Angles and Triangles*

"*Can you name the shortest word in the English language with all five vowels (not necessarily in order)? The word is seven letters long.*"

**Before** I wrote this chapter, I formulated in my head how the chapter would go. Too many of the questions on angles had to do with triangles. So I decided to write the chapters together. Let's start with some definitions.

## TYPES OF ANGLES

There are several ways to classify angles, such as by angle measure, as shown here:

**Acute angle:** An angle of less than 90°.

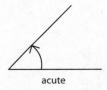

acute

**Right angle:** A 90° angle. As we will see, some other words that indicate a right angle or angles are perpendicular (⊥), altitude, and height.

right

**Obtuse angle:** An angle of more than 90° but less than 180°.

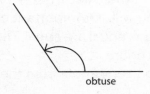

obtuse

**Straight angle:** An angle of 180°.

Reflex angle:** An angle of more than 180° but less than 360°.

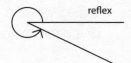

Angles are also named for their relation to other angles, such as:

**Supplementary angles:** Two angles that total 180°. $\angle 1 + \angle 2 = 180°$.

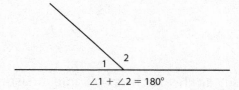

$\angle 1 + \angle 2 = 180°$

**Complementary angles:** Two angles that total 90°. $\angle 1 + \angle 2 = 90°$.

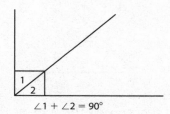

$\angle 1 + \angle 2 = 90°$

You probably learned that angles are congruent and measures of angles are equal. I am using what I learned; it is simpler and makes understanding easier. So "angle 1 equals angle 2" (or $\angle 1 = \angle 2$) means the angles are both congruent and equal in degrees.

## ANGLES FORMED BY PARALLEL LINES

Let's look at angles formed when a line crosses two parallel lines. In the next figure, $\ell_1 \| \ell_2$, and $t$ is a **transversal**, a line that cuts two or more lines. It is not important that you know the names of these angles, although many of you will. It is important only to know that angles formed by a line crossing parallel lines that look equal are equal. The angles that are adjacent to these angles add to 180° with them. In this figure, $\angle 1 = \angle 4 = \angle 5 = \angle 8$ and $\angle 2 = \angle 3 = \angle 6 = \angle 7$. Any angle from the first group added to any angle from the second group totals 180°.

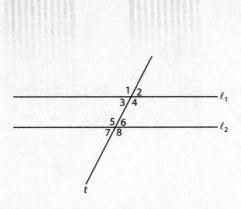

**Vertical angles**, which are the opposite angles formed when two lines cross, are equal. In the figure below, $\angle 1 = \angle 3$ and $\angle 2 = \angle 4$. Also, $\angle 1 + \angle 2 = \angle 2 + \angle 3 = \angle 3 + \angle 4 = \angle 4 + \angle 1 = 180°$.

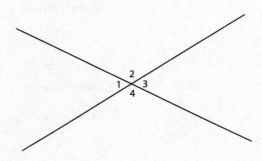

**(Q)** **Let's do some multiple-choice exercises.**

**Exercise 1:** $\angle b =$

A. 45°          D. 105°

B. 60°          E. 135°

C. 90°

**Exercise 2:** $\ell_1 \parallel \ell_2.\ m - n =$

A. 30°          D. 90°

B. 50°          E. 180°

C. 65°

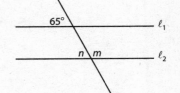

**Exercise 3:** $y + z =$

A. $180° - x$          D. $90° + \dfrac{5x}{4}$

B. $180° - \dfrac{x}{4}$          E. $90° - \dfrac{5x}{4}$

C. $45° - \dfrac{x}{4}$

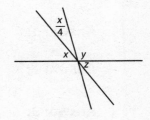

**Exercise 4:** $180° - w =$

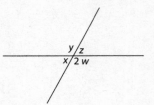

    **A.** $x + w$          **D.** $y - z$

    **B.** $x + y$          **E.** $z - w$

    **C.** $y + z$

**Exercise 5:** $b =$

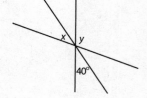

    **A.** $5.5°$          **D.** $12.5°$

    **B.** $7°$            **E.** Cannot be

    **C.** $10°$               determined

**Exercise 6:** $y$ (in terms of $x$) $=$

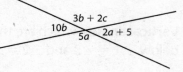

    **A.** $x$             **D.** $140° + x$

    **B.** $x + 40°$      **E.** $320° - x$

    **C.** $140° - x$

**Exercise 7:** $\angle x =$

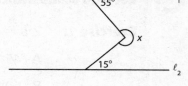

    **A.** $70°$          **D.** $290°$

    **B.** $110°$         **E.** $345°$

    **C.** $210°$

**Exercise 8:** The ratio of $a°$ to $(a + b)°$ is 3 to 8; $a =$

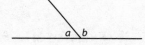

    **A.** $60°$          **D.** $108°$

    **B.** $67.5°$        **E.** $112.5°$

    **C.** $72°$

**(A)** **Now let's look at the answers.**

**Answer 1:**   A: $3a + a = 180°$; $a = 45°$ and $b = a = 45°$.

**Answer 2:**   B: $n = 65°$ and $n + m = 180°$; so $m = 115°$, and $m - n = 50°$.

**Answer 3:**   B: $\dfrac{x}{4} + y + z = 180°$, so $y + z = 180° - \dfrac{x}{4}$.

**Answer 4:**   A: Below the line, $x + 2w = x + w + w = 180°$, so $x + w = 180° - w$.

**Answer 5:** **A:** This is a toughie. Don't look at vertical angles, look at the supplementary angles. On the bottom, we have $5a + 2a + 5° = 180°$, so $7a = 175°$, and $a = 25°$. Then, on the left, $10b + 5a = 180°$. Substituting $a = 25°$, we get $10b = 180° − 125° = 55°$, or $b = 5.5°$.

If we had looked at the vertical angles after we determined that $a = 25°$, then $10b = 2a + 5° = 2(25°) + 5° = 55°$, so $b = 5.5°$.

**Answer 6:** **C:** $x + y + 40° = 180°$; so $y = 140° − x$.

**Answer 7:** **D:** Draw $\ell_3 \parallel \ell_1$ and $\ell_2$.

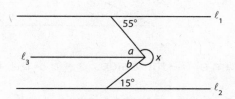

Then $\angle b = 15°$ and $\angle a = 55°$, so

$\angle x = 360° − (15° + 55°) = 290°$.

**Answer 8:** **B:** One way to answer this exercise is to say $a = \left(\dfrac{3}{8}\right) \times 180$. You will

notice that if you divide 8 into 180, you will have a fraction (or a decimal).

So only answer choices B or E could be correct. Because B is $< 90$, B must

be the correct answer. Or if you recognize that $(a + b)° = 180°$, you could

use the ratio $\dfrac{a}{180} = \dfrac{3}{8}$, which gives $67.5°$.

# TRIANGLES

## Basics about Triangles

A **triangle** is a polygon with three sides. Angles are usually indicated with capital letters. The side opposite the angle is indicated with the same letter, only lowercase.

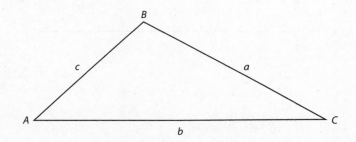

You should know the following general facts about triangles.

- The **sum of the angles** of a triangle is 180°.

- The **altitude**, or **height** (h), of △ABC shown below is the line segment drawn from a vertex perpendicular to the base, extended if necessary. The **base** of each triangle is AC = b.

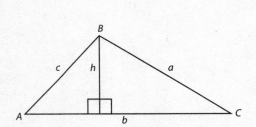 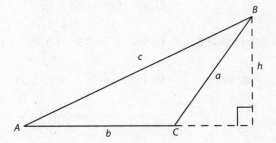

- The **perimeter of a triangle** is the sum of the three sides: $p = a + b + c$.

- The **area of a triangle** is $A = \frac{1}{2}bh$. The reason is that a triangle is half a rectangle. Because the area of a rectangle is base times height, the area of a triangle is half that of a rectangle, as shown in the figure below.

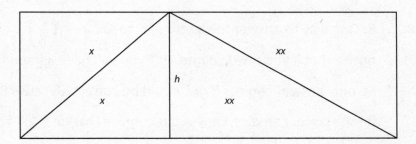

- An **angle bisector** is a line that bisects an angle in a triangle. In the figure below, BD bisects ∠ABC if ∠1 = ∠2.

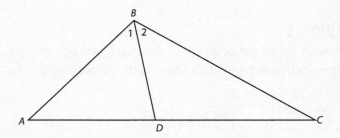

- A **median** is a line drawn from any angle of a triangle to the midpoint of the opposite side. In the figure below, *BD* is a median to side *AC* if *D* is the midpoint of *AC*.

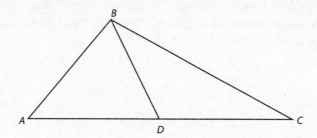

There are many kinds of triangles. One way to describe them is by their sides.

- A **scalene** triangle has three unequal sides and three unequal angles.

- An **isosceles** triangle has at least two equal sides. In the figure below, side *BC* (or *a*) is called the **base**; it may be equal to, greater than, or less than any other side. The **legs**, *AB* = *AC* (or *b* = *c*) are equal. Angle *A* is the **vertex angle**; it may equal the others, or be greater than or less than the others. The **base angles** are always equal: ∠*B* = ∠*C*.

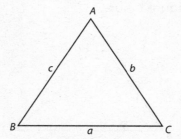

- An **equilateral** triangle is a triangle with all equal sides. All angles equal 60°, so this triangle is sometimes called an **equiangular** triangle. For an equilateral triangle of side *s*, the perimeter $p = 3s$, and the area $A = \dfrac{s^2\sqrt{3}}{4}$.

Triangles can also be described by their angles.

- An **acute** triangle has three angles that are less than 90°.

- A **right** triangle has one right angle, as shown in the figure below. The **right angle** is usually denoted by the capital letter *C*. The **hypotenuse** *AB* is the side opposite the right angle. The **legs**, *AC* and *BC*, are not necessarily equal. ∠*A* and ∠*B* are always **acute** angles.

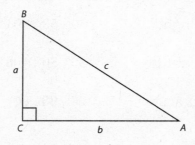

- An **obtuse** triangle has one angle between 90° and 180°.

An **exterior angle** of a triangle is formed by extending one side. In the figure below, $\angle 1$ is an exterior angle. An exterior angle equals the sum of the other two angles (called its two remote interior angles): $\angle 1 = \angle A + \angle B$.

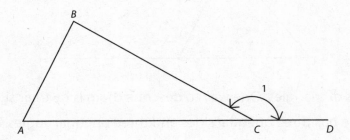

There are three other facts about triangles you should know:

1.  The sum of any two sides of a triangle must be greater than the third side.

2.  The largest side lies opposite the largest angle, and the largest angle lies opposite the largest side, as shown in the figure below.

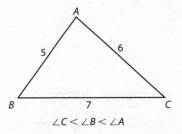

$$\angle C < \angle B < \angle A$$

3.  A line segment joining the midpoints of two sides of a triangle is parallel to the third side and equals half of the third side.

**Example 1:** Give one set of angles for a triangle that satisfies the following descriptions:

| Description | Solution |
| --- | --- |
| **a.** Scalene, acute | 50°, 60°, 70° |
| **b.** Scalene, right | 30°, 60°, 90°. We will deal with this one soon. |
| **c.** Scalene, obtuse | 30°, 50°, 100° |
| **d.** Isosceles, acute | 20°, 80°, 80° |

**e.** Isosceles, right    Only one answer: 45°, 45°, 90°. We will also deal with this one soon.

**f.** Isosceles, obtuse    20°, 20°, 140°

**g.** Equilateral    Only one answer: three 60° angles

 **Note** *The solutions to Example 1 are not necessarily the only possible answers.*

Let's first do some exercises with angles. Then we'll turn to area and perimeter exercises. We'll finish the chapter with our famous friend Pythagoras and his well-known theorem.

**Ⓠ** **Let's do some more multiple-choice exercises.**

**Exercise 9:** Two sides of a triangle are 4 and 7. If only integer measures are allowed for the sides, the third side must be taken from which set?

**A.** {5, 6, 7, 8, 9, 10, 11}     **D.** {3, 4, 5, 6, 7, 8, 9, 10, 11}

**B.** {4, 5, 6, 7, 8, 9, 10}     **E.** {1, 2, 3, 4, 5, 6, 7, 8, 9, 10, 11}

**C.** {3, 4, 5, 6, 7, 8, 9, 10}

**Exercise 10:** Arrange the sides in order, largest to smallest, for the figure shown at right.

**A.** $a > b > c$     **D.** $b > c > a$

**B.** $a > c > b$     **E.** $c > a > b$

**C.** $b > a > c$

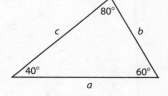

**Exercise 11:** $\angle TVW = 10x$; $x$ could be

**A.** 3°     **D.** 16°

**B.** 6°     **E.** 20°

**C.** 9°

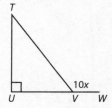

**Exercise 12:** Write $b$ in terms of $a$:

**A.** $a + 90°$     **D.** $180° - a$

**B.** $2a$     **E.** $180° - 2a$

**C.** $2a + 90°$

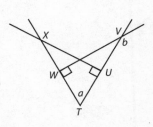

**Exercise 13:**  The largest angle is

       **A.** 30°        **D.** 80°

       **B.** 50°        **E.** 90°

       **C.** 70°

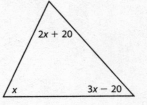

**Exercise 14:**  $\ell_1 \parallel AB$; $y =$

       **A.** 40°        **D.** 80°

       **B.** 60°        **E.**  Cannot be

       **C.** 70°                determined

Use $\triangle ABC$ for Exercises 15 and 16.

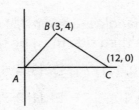

**Exercise 15:**  The area of $\triangle ABC$ is

       **A.** 18        **D.** 48

       **B.** 24        **E.** 60

       **C.** 36

**Exercise 16:**  The perimeter of $\triangle ABC$ is

       **A.** $17 + \sqrt{97}$        **D.** $\sqrt{266}$

       **B.** 27        **E.** $10\sqrt{10}$

       **C.** 32

For Exercises 17 and 18, use this figure of a square with an equilateral triangle on top of it, $AE = 20$.

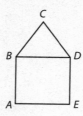

**Exercise 17:**   The perimeter of $ABCDE$ is

      **A.** 50            **D.** 160

      **B.** 100          **E.** 200

      **C.** 120

**Exercise 18:**   The area of $ABCDE$ is

      **A.** 600          **D.** 800

      **B.** $100(4 + \sqrt{2})$       **E.** 1,000

      **C.** $100(4 + \sqrt{3})$

For Exercises 19 and 20, use $\triangle ABC$ with midpoints $X$, $Y$, and $Z$.

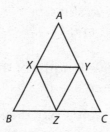

**Exercise 19:**   If the perimeter of $\triangle ABC$ is 1, the perimeter of $\triangle XYZ$ is

      **A.** $\dfrac{1}{16}$          **D.** $\dfrac{1}{2}$

      **B.** $\dfrac{1}{8}$          **E.** 1

      **C.** $\dfrac{1}{4}$

**Exercise 20:**  If the area of $\triangle ABC$ is 1, the area of $\triangle XYZ$ is

A. $\dfrac{1}{16}$    D. $\dfrac{1}{2}$

B. $\dfrac{1}{8}$    E. 1

C. $\dfrac{1}{4}$

**Exercise 21:**  In the figure shown, $BC = \dfrac{1}{3}BD$. If the area of $\triangle ABC = 10$, the area of rectangle $ABDE$ is

A. 30    D. 120

B. 40    E. Cannot be
determined

C. 60

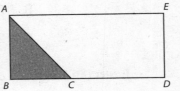

 A **Let's look at the answers.**

**Answer 9:**  B: The third side $s$ must be greater than the difference and less than the sum of the other two sides, or $> 7 - 4$ and $< 7 + 4$. Thus the third side must be between 3 and 11.

**Answer 10:**  B: Judge the relative lengths of the sides by the size of the angles opposite them. Then $a > c > b$.

Watch out for the words "Not drawn to scale." If it is a simple figure, "not drawn to scale" usually means it is not drawn to scale, and you cannot assume relative sizes without being given actual measurements. However, if it is a semi-complicated or complicated figure, the figure probably *is* drawn to scale.

**Answer 11:**  D: $\angle TVW$ must be between $90°$ and $180°$, so $9° < x < 18°$.

**Answer 12:**  A: This is really tricky. $UX$ is drawn to confuse you. In $\triangle TVW$, $b$ is the exterior angle, so $b = a + 90°$.

**Answer 13:**  D: $x + 2x + 20 + 3x - 20 = 180$, or $6x = 180$, so $x = 30$. $2x + 20 = 80$ and $3x - 20 = 70$. The largest angle is $80°$.

**Answer 14:**  A: $2x + x + 60 = 180$; $x = 40°$. But $y = x = 40°$ (because $\ell_1 \parallel AB$).

**Answer 15:**   **B:** $A = \dfrac{1}{2}bh = \dfrac{1}{2} \times 12 \times 4 = 24$.

**Answer 16:**   **A:** Use the distance formula to find sides $AB$ and $BC$. $p = AC + AB + BC = 12 + \sqrt{4^2 + 3^2} + \sqrt{(3-12)^2 + (4-0)^2} = 12 + \sqrt{25} + \sqrt{97} = 17 + \sqrt{97}$.

**Answer 17:**   **B:** Do not include $BD$; $p = 5 \times 20 = 100$.

**Answer 18:**   **C:** Area $= s^2 + \dfrac{s^2\sqrt{3}}{4} = 20^2 + \dfrac{20^2\sqrt{3}}{4} = 400 + 100\sqrt{3} = 100\,(4 + \sqrt{3})$

**Answer 19:**   **D:** All the sides of $\triangle XYZ$ are medians of $\triangle ABC$. If the perimeter of $\triangle ABC$ is 1, then all the sides of $\triangle XYZ$ are half of those of $\triangle ABC$, and so is the perimeter.

**Answer 20:**   **C:** If the sides of $\triangle XYZ$ are half of those of $\triangle ABC$, the area of $\triangle XYZ$ is $\left(\dfrac{1}{2}\right)^2 A = \dfrac{1}{4}A = \dfrac{1}{4}$.

**Answer 21:**   **C:** If we draw lines parallel to $DE$ to divide the original rectangle into three congruent rectangles, and then divide each rectangle into two triangles, we see that each triangle is one-sixth of the rectangle. So the area of the rectangle is $6(10) = 60$.

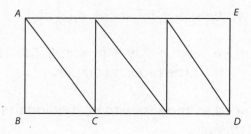

We'll have more of these type of exercises as part of Chapter 12, where we discuss circles.

Let's go on to good old Pythagoras.

## PYTHAGOREAN THEOREM

This is perhaps the most famous math theorem of all. Most theorems have one proof. A small fraction have two. This theorem, however, has more than a hundred, including three by past presidents of the United States. We've had some smart presidents who actually knew some math.

The Pythagorean Theorem simply states:

*In a right triangle, the hypotenuse squared is equal to the sum of the squares of the legs.*

In symbols,

$$c^2 = a^2 + b^2.$$

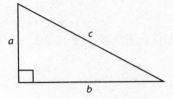

As a teacher, I must show you one proof.

*Proof:*

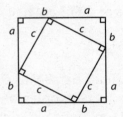

In this figure, the area of the larger square equals that of the smaller square plus the four congruent triangles. In symbols, $(a + b)^2 = c^2 + 4\left(\frac{1}{2}ab\right)$.

Multiplying this equation out, we get $a^2 + 2ab + b^2 = c^2 + 2ab$. Then subtracting $2ab$ from both sides, we get $c^2 = a^2 + b^2$. The proof is complete.

There are two basic problems you need to know how to do: finding the hypotenuse and finding one of the legs of the right triangle.

**Example 2:**   Solve for $x$:

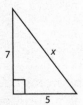

**Solution:**   $x^2 = 7^2 + 5^2; x = \sqrt{74}$.

**Example 3:** Solve for $x$:

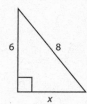

**Solution:** $8^2 = 6^2 + x^2$, or $x^2 = 64 - 36 = 28$. So $x = \sqrt{28} = \sqrt{2 \times 2 \times 7} = 2\sqrt{7}$.

Notice that the hypotenuse squared is always by itself, whether it is a number or a letter.

## Pythagorean Triples

It is a good idea to memorize these triples. Those in the 3-4-5 and 5-12-13 groups are just multiples of the groups. Below are the sides of the most important right triangles. The hypotenuse is always listed third in each group.

The 3-4-5 group:          3-4-5, 6-8-10, 9-12-15, 12-16-20, 15-20-25

The 5-12-13 group:       5-12-13, 10-24-26

The rest:                      8-15-17, 7-24-25, 20-21-29, 9-40-41, 11-60-61

## Special Right Triangles

You ought to know two other special right triangles, the isosceles right triangle (with angles 45°-45°-90°) and the 30°-60°-90° right triangle. The facts about these triangles can all be found by using the Pythagorean Theorem.

1.   The 45°-45°-90° isosceles right triangle:

- The legs are equal.
- To find a leg given the hypotenuse, divide the hypotenuse by $\sqrt{2}$ (or multiply by $\frac{\sqrt{2}}{2}$).
- To find the hypotenuse given a leg, multiply the leg by $\sqrt{2}$.

**Example 4:** Find $x$ and $y$ for this isosceles right triangle.

**Solution:** $x = 5$ (the legs are equal); $y = 5\sqrt{2}$.

**Example 5:** Find x and y for this isosceles right triangle.

**Solution:** $x = y = \dfrac{18}{\sqrt{2}} = 18 \times \dfrac{\sqrt{2}}{2} = 9\sqrt{2}$.

2. The 30°-60°-90° right triangle.

- If the shorter leg (opposite the 30° angle) is not given, get it first. It is always half the hypotenuse.

- To find the short leg given the hypotenuse: divide the hypotenuse by 2.

- To find the hypotenuse given the short leg: multiply the short leg by 2.

- To find the short leg given the long leg: divide the long leg by $\sqrt{3}$ (or multiply by $\dfrac{\sqrt{3}}{3}$).

- To find the long leg given the short leg: multiply the short leg by $\sqrt{3}$.

**Example 6:** Find x and y for this right triangle.

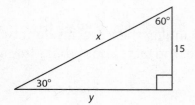

**Solution:** The short leg is given (15); $x = 2(15) = 30$; $y = 15\sqrt{3}$.

**Example 7:** Find x and y for this right triangle.

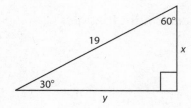

**Solution:** $x = \dfrac{19}{2} = 9.5$; $y = 9.5\sqrt{3}$.

**Example 8:**    Find *x* and *y* for this right triangle.

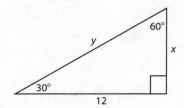

**Solution:**    $x = \dfrac{12}{\sqrt{3}} = 12\dfrac{\sqrt{3}}{3} = 4\sqrt{3}$; $y = 2(4\sqrt{3}) = 8\sqrt{3}$.

**Q**    **Let's do a few more multiple-choice exercises.**

**Exercise 22:**    Two sides of a right triangle are 3 and $\sqrt{5}$.

    **I:**    The third side is 2.

    **II:**    The third side is 4.

    **III:**    The third side is $\sqrt{14}$.

    Which of the following choices is correct?

    **A.** Only II is true.      **D.** Only I and III are true.

    **B.** Only III is true.      **E.** I, II, and III are true.

    **C.** Only I and II are true.

**Exercise 23:**    The area of square *ABCD* =

    **A.** 50           **D.** 576

    **B.** 100        **E.** 625

    **C.** 225

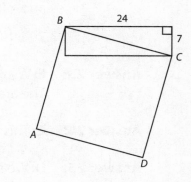

**Exercise 24:**    *x* =

    **A.** 16          **D.** 22

    **B.** 18          **E.** 24

    **C.** 20

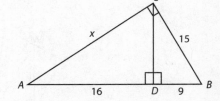

**Exercise 25:**   $c^2 - b^2 =$

A. 72

D. 194

B. 144

E. 288

C. 216

**Exercise 26:**   $x =$

A. 1

D. 4

B. 2

E. 4.5

C. 3

**Exercise 27:**   A 25-foot ladder is leaning on the floor. Its base is 15 feet from the wall. If the ladder is pushed until it is only 7 feet from the wall, how much farther up the wall is the ladder pushed?

A. 4 feet

D. 20 feet

B. 8 feet

E. 24 feet

C. 12 feet

   **Let's look at the answers.**

**Answer 22:**   D: Try the Pythagorean Theorem with various combinations of 3, $\sqrt{5}$, and $x$ (the third side). The only ones that work are Statement I:

$2^2 + \left(\sqrt{5}\right)^2 = 3^2$; and Statement III: $3^2 + \left(\sqrt{5}\right)^2 = \left(\sqrt{14}\right)^2$.

**Answer 23:**   E: We recognize the right triangle as a 7-24-25 triple, so side $BC = 25$. The area of the square is $(25)^2 = 625$.

**Answer 24:**   C: This is a 15-20-25 triple, so $x = 20$.

**Answer 25:**   D: We see that $MN$ is the side of two triangles. By the Pythagorean Theorem, we get $c^2 - b^2 = x^2 + y^2 = \left(5\sqrt{2}\right)^2 + 12^2 = 50 + 144 = 194$.

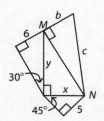

**Answer 26:**  **D:** This triangle is a 12-16-20 triple, so $3x + 2x = 5x = 20$, and $x = 4$.

**Answer 27:**  **A:** The first figure is a 15-20-25 right triangle with the ladder 20 feet up the wall. The second figure is a 7-24-25 triple with the ladder 24 feet up the wall. The ladder is pushed another $24 - 20 = 4$ feet up the wall.

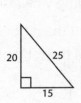 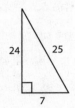

## SIMILAR TRIANGLES

Two triangles are similar if the corresponding angles are congruent and corresponding sides are in proportion.

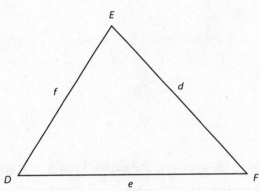 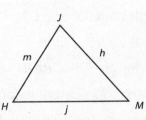

In the above figure, this means $\angle D \cong \angle H$, $\angle E \cong \angle J$, and $\angle F \cong \angle M$ and $\dfrac{f}{m} = \dfrac{d}{h} = \dfrac{e}{j}$.

$\dfrac{f}{m} = \dfrac{d}{h} = \dfrac{e}{j}$ is called the **ratio of similarity**.

**Example 9:**  For the figures below,

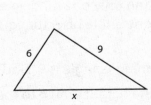 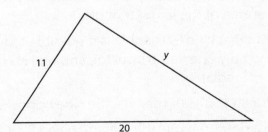

a. What is the ratio of similarity?

b. Find $x$ and $y$.

c. What is the ratio of their areas?

**Solutions:**    a.  The left sides in each triangle are given. The ratio of similarity is $\dfrac{6}{11}$.

b.  $\dfrac{6}{11} = \dfrac{x}{20}$; $11x = 120$; $x = \dfrac{120}{11}$; $\dfrac{6}{11} = \dfrac{9}{y}$; $6y = 99$; $y = \dfrac{99}{6}$ or $\dfrac{33}{2}$. Once you start with the length of the smallest side of the first triangle on the top of a fraction, the corresponding smallest side of the second triangle must always be on the top of the ratio.

c.  The ratio of the areas is the square of the ratio of similarity. In this case, the ratio of the areas would be $\dfrac{36}{121}$.

 *For three-dimensional similar figures, the ratio of their volumes would be the cube of the ratio of similarity.*

## Chapter 10 Quiz

For questions 1–4, describe the triangle with regard to angles and sides.

1.  Two angles are 20° and 30°.

2.  Two angles are 60° and 60°.

3.  Three sides are 7, 7, and 6.

4.  Three sides are 9, 40, are 41.

5.  One of two supplementary angles is 3 more than twice the other. Find the angles.

6.  The base of an isosceles triangle is 6 more than a leg. The perimeter is 99 feet. Find the sides.

7.  The hypotenuse of a right triangle is 8 and one leg is 7; find the area and perimeter of the triangle.

8.  The area of an equilateral triangle is $100\sqrt{3}$. Find its perimeter.

9.  The ratio of similarity of two triangles is 2 to 3. If 24 is the area of the smaller triangle, find the area of the larger triangle.

10. For what kind of triangle is it possible that one and only one median equals one and only one angle bisector, and one and only one angle bisector equals one and only one median?

11. In a 30°, 60°, 90° triangle, the side opposite the middle angle is 9. Find its area.

12. In a right triangle, the legs are 6 and 8. What is the height drawn to the hypotenuse?

13. A regular hexagon consists of six equilateral triangles. One side is 4; find the perimeter and area.

14. An isosceles triangle has legs of 15. If the base is 24, find its area.

15. If two vertical angles are $7x - 3$ and $2x + 17$, what are their measures?

16. Two interior angles on the same side of a transversal cutting two parallel lines are $3x + 5$ and $5x + 15$. Find the angles.

17. An isosceles right triangle has an area of 50. Find the hypotenuse.

18. Find the side of an equilateral triangle if a square of side 25 and the triangle have the same perimeter.

19. The second side of a triangle is 3 less than the first, and the third side is 9 less than the second. If the perimeter is 90, find the sides.

20. If the area of a square is equal to the area of an isosceles right triangle, write the leg of the right triangle in terms of $s$, the side of the square.

## Answers to Chapter 10 Quiz

1. The third angle is $180° - 50° = 130°$; obtuse and scalene.

2. The third angle is also $60°$; equilateral and equiangular (acute also OK).

3. Isosceles and acute.

4. Right and scalene.

5. $x + 2x + 3 = 180$; $x = 59°$; $2x + 3 = 121°$.

6. $x = $ leg; $x + x + x + 6 = 99$; $x = 31$; the three sides are 31, 31, and 37.

7. The other leg $= \sqrt{8^2 - 7^2} = \sqrt{15}$; $A = \frac{1}{2}(\ell_1 \times \ell_2) = \frac{7\sqrt{15}}{2}$; the perimeter is $15 + \sqrt{15}$.

8. $A = \frac{s^2\sqrt{3}}{4} = 100\sqrt{3}$. So $s = 20$ and $p = 60$.

9. The ratio of the areas is $\frac{4}{9}$; $\frac{4}{9} = \frac{24}{x}$; $x = 54$.

10. Isosceles triangle from the vertex to the base or an equilateral triangle.

11. The short leg is $\frac{9}{\sqrt{3}} = 3\sqrt{3}$; $A = \frac{1}{2}b \times h = \frac{1}{2}9(3\sqrt{3}) = 13.5\sqrt{3}$.

12. The hypotenuse is 10; $A = \frac{1}{2}b_1h_1 = \frac{1}{2}b_2h_2$; substitute the values and multiply both sides by 2 to get $(6)(8) = 10(h_2)$; $h_2 = 4.8$.

13. $p = 6(4) = 24$; $A = 6 \times \frac{s^2\sqrt{3}}{4} = 24\sqrt{3}$.

14. The height from the vertex bisects the base; we have a 9, 12, 15 triangle; the height is 9; $A = \frac{1}{2}(bh) = \frac{1}{2}(24)(9) = 108$.

15. $7x - 3 = 2x + 17$; $x = 4°$; the vertical angles are $25°$ each.

16. $3x + 5 + 5x + 15 = 180$; $x = 20°$; the angles are 65° and 115°.

17. $A = \frac{1}{2}\ell^2 = 50$; $\ell^2 = 100$; $\ell = 10$; $h = \ell\sqrt{2} = 10\sqrt{2}$.

18. $4(25) = 3s$; $s = 33\frac{1}{3}$.

19. $x + (x - 3) + (x - 12) = 90$; $3x - 15 = 90$; $x = 35$; the answers are 35, 32, and 23.

20. If the areas are the same, $s^2 = \frac{1}{2}\ell^2$, and $\ell = s\sqrt{2}$.

That's all for angles and triangles for now. We will see more when circles are discussed in Chapter 12. For now, though, let's look at rectangles and other polygons.

**Answer to "Bob Asks":** Sequoia

# CHAPTER 11: *Other Two-Dimensional Figures*

"*Sue had a 20% increase, followed by a 20% decrease. Lou had a 20% decrease, followed by a 20% increase. Who got the better deal? What was the difference increase or decrease between Sue and Lou, or was it the same?*"

**We** now deal with the rest of the polygons (closed figures with line-segment sides).

## QUADRILATERALS

### Parallelograms

A **parallelogram** is a **quadrilateral** (four-sided polygon) with parallel opposite sides.

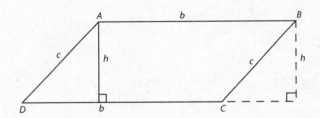

You should know the following properties about parallelograms:

*   The opposite angles are equal. $\angle DAB = \angle BCD$ and $\angle ADC = \angle ABC$.
*   The consecutive angles are supplementary. $\angle DAB + \angle ABC = \angle ABC + \angle BCD = \angle BCD + \angle CDA = \angle CDA + \angle DAB = 180°$.
*   The opposite sides are equal. $AB = CD$ and $AD = BC$.
*   Area $= A = bh$.
*   Perimeter $= p = 2b + 2c$.
*   The diagonals bisect each other. $AE = EC$ and $DE = EB$ in the figure below.

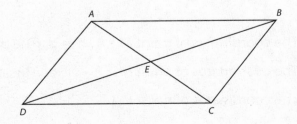

**Example 1:**    For parallelogram *RSTU* shown in the figure below, find the following if *RU* = 10:

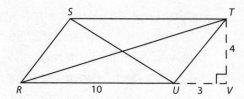

a. The area                 c. Diagonal *RT*

b. The perimeter            d. Diagonal *SU*

**Solutions:**    a. $A = bh = (10)(4) = 40$ square units. The whole test should be this easy!

b. We have to find the length of *RS* = *TU*. *TU* = 5 because it is the hypotenuse of a 3-4-5 right triangle. So the perimeter is $p = 2(10) + 2(5) = 30$ units.

c. $RT = \sqrt{(RV)^2 + (TV)^2} = \sqrt{13^2 + 4^2} = \sqrt{185}$

d.

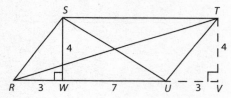

To find diagonal *SU*, draw the other altitude *SW* as pictured.

$$SU = \sqrt{WU^2 + SW^2} = \sqrt{7^2 + 4^2} = \sqrt{65}$$

**Example 2:**    For parallelogram *WXYZ* with altitudes *XM* and *YN*, find the following in terms of *a*, *b*, and *c*:

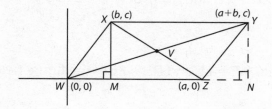

a. The coordinates of point *M*        d. The perimeter

b. The coordinates of point *N*        e. The area

c. The coordinates of point *V*

**Solutions:**

a. $M$ has the same $x$-coordinate as point $X$ and the same $y$-coordinate as point $W$, so the coordinates of $M$ are $(b, 0)$.

b. $N$ has the same $x$-coordinate as point $Y$ and the same $y$-coordinate as point $W$, so the coordinates of $N$ are $(a + b, 0)$.

c. $V$ is halfway between $W$ and $Y$, so use the formula for the midpoint between $Y(a + b, c)$ and $W(0, 0)$: Midpoint $V = \left( \dfrac{(a + b) + 0}{2}, \dfrac{c + 0}{2} \right) = \left( \dfrac{a + b}{2}, \dfrac{c}{2} \right)$.

d. $WZ = XY$ is length $a$. By the distance formula, $WX = ZY = \sqrt{(b - 0)^2 + (c - 0)^2} = \sqrt{b^2 + c^2}$. Therefore, the perimeter is $p = 2a + 2\sqrt{b^2 + c^2}$.

e. Area $= A =$ base $\times$ height $= ac$.

**Example 3:** For parallelogram $EFGH$, find the smaller angle.

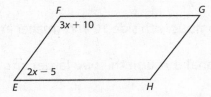

**Solution:** Consecutive angles of a parallelogram are supplementary. Therefore, $(3x+10)° + (2x - 5)° = 180°$; $x = 35°$; so the smaller angle is $2(35°) - 5° = 65°$. Be careful to give the answer the COMPASS wants. Two other, but incorrect, answer choices would be $35°$ and $115°$, for those who do not read carefully!!!

## Rhombus

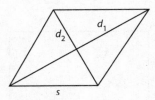

A **rhombus** is an equilateral parallelogram.

Thus, a rhombus has all of the properties of a parallelogram plus the following:

* All sides are equal.
* The diagonals are perpendicular bisectors of each other.

- Perimeter $= p = 4s$.
- Area $= A = bh = \dfrac{1}{2} \times d_1 \times d_2$, the area equals half the product of its diagonals.

**Example 4:**   For the rhombus given below with side $s = 13$ and larger diagonal $BD = 24$, find the other diagonal and the area.

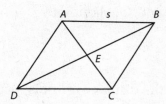

**Solution:**   $AB = 13$ and $BD = 24$. Because the diagonals bisect each other, $BE = 12$.

The diagonals are perpendicular to each other, so $\triangle ABE$ is a 5-12-13 right

triangle, and $AE = 5$. Therefore, the other diagonal $AC = 10$. The area is

$A = \dfrac{1}{2} \times BD \times AC = \dfrac{1}{2}(24)(10) = 120$ square units.

**Example 5:**   Find the area of a rhombus with side 10 and smaller interior angle of 60°.

**Solution:**   If we draw the diagonal through the two larger angles, we will have two

congruent equilateral triangles. The area of this rhombus is twice the

area of each triangle, or $2 \times \dfrac{s^2\sqrt{3}}{4}$. Because $s = 10$, the area is

$A = 2 \times \dfrac{10^2\sqrt{3}}{4} = 50\sqrt{3}$ square units.

Now let's go on to more familiar territory.

## Rectangle

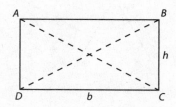

A **rectangle** is a parallelogram with right angles.

Therefore, it has all of the properties of a parallelogram plus the following:

- All angles are 90°.

- Diagonals are equal (but *not* perpendicular).

- Perimeter $= p = 2b + 2h$.

- Area $= A = bh$. This is a **postulate** (law taken to be true without proof) from which we get the area of all other figures with sides that are line segments.

**Example 6:** One base of a rectangle is 8, and one diagonal is 9. Find all the sides and the other diagonal. Find the perimeter and area.

**Solution:** The top base and bottom base are both 8. Both diagonals are 9. The other two sides are each $\sqrt{9^2 - 8^2} = \sqrt{17}$. So the perimeter is $p = 16 + 2\sqrt{17}$ units; and the area is $A = 8\sqrt{17}$ square units.

**Example 7:** $AB = 10$, $BC = 8$, $EF = 6$, and $FG = 3$. Find the area of the shaded region of the figure below.

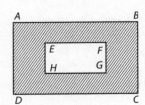

**Solution:** The area of the shaded region is the area of the outside rectangle minus the area of the inside one. $A = (10)(8) - (6)(3) = 62$ square units.

**Example 8:** In polygon $ABCDEF$, $BC = 30$, $AF = 18$, $AB = 20$, and $CD = 11$. Find the perimeter and area of the polygon.

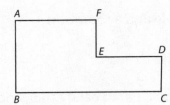

**Solution:** Draw a line through $DE$, hitting $AB$ at point $G$. Then $AF = GE$ and $BC = GD$. Because $DG = 30$ and $GE = 18$, $DE = 12$. $AB = CD + EF$. $AB = 20$ and $CD = 11$, so $EF = 9$. This gives the lengths of all the sides. The perimeter, thus, is $p = AB + BC + CD + DE + EF + AF = 20 + 30 + 11 + 12 + 9 + 18 = 100$ units.

The area of rectangle $BCDG$ is $BC \times CD = (30)(11) = 330$. The area of rectangle $AFEG$ is $AF \times FE = (18)(9) = 162$. Therefore, the total area is $330 + 162 = 492$ square units. There are other ways to find this area, as you might be able to see.

**Example 9:**  The areas of the pictured rectangle and triangle are the same.
If $\dfrac{LW}{4} = 20$, what is $bh$?

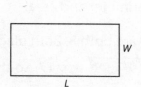 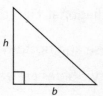

**Solution:**  $\dfrac{LW}{4} = 20$; so $LW = 80$, which is the area of the rectangle. Because that also is the area of the triangle, $\dfrac{1}{2}bh = 80$. So $bh = 160$.

## Square

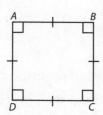

A **square** is a rectangle with equal sides, or it can be thought of as a rhombus with four equal 90° angles.

Therefore, it has all of the properties of a rectangle and a rhombus:

- All sides are equal.
- All angles are 90°.
- Both diagonals bisect each other, are perpendicular to each other, and are equal.
- Each diagonal $d = d_1 = d_2 = s\sqrt{2}$, where $s =$ a side.
- Perimeter $= p = 4s$.
- Area $= A = \dfrac{d^2}{2} = s^2$.

**Example 10:** What is the area of this square?

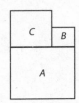

**Solution:** Because it is a square, $5x - 1 = x + 1$, so $x = \dfrac{1}{2}$ and $x + 1 = 1\dfrac{1}{2}$. Then $A = \left(1\dfrac{1}{2}\right)^2 = 2\dfrac{1}{4}$.

**Example 11:** The area of square $C$ is 36; the area of square $B$ is 25. What is the area of square $A$?

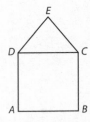

**Solution:** The side of square $C$ must be 6, and the side of square $B$ must be 5.

Therefore, the side of square $A$ is 11, and the area of square $A$ is $11^2 = 121$.

Use the figure below for Examples 12 and 13. It is a square surmounted by an equilateral triangle (I've always wanted to write that word). $AB = 10$.

**Example 12:** What is the perimeter of the figure?

**Solution:** The perimeter is $5(10) = 50$. Note that $CD$ is *not* part of the perimeter.

**Example 13:** What is the area of the figure?

**Solution:** $A = s^2 + \dfrac{s^2\sqrt{3}}{4} = 10^2 + \dfrac{10^2\sqrt{3}}{4} = 100 + 25\sqrt{3} = 25(4 + \sqrt{3})$.

## Trapezoid

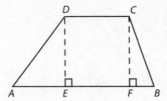

A **trapezoid** is a quadrilateral with exactly one pair of parallel sides.

Because a trapezoid is *not* a type of parallelogram, it has its own unique set of properties, as follows:

* The parallel sides, *AB* and *CD*, are called **bases**.

* The heights, *DE* and *CF*, are equal.

* The legs, *AD* and *BC*, may or may not be equal.

* The diagonals, *AC* and *BD*, may or may not be equal.

* Perimeter $= p = AB + BC + CD + DA$.

* Area $= A = \frac{1}{2}h(b_1 + b_2)$, where $b_1$ and $b_2$ are the bases and $h$ is the height.

> **Note**   *If we draw one of the diagonals, we see that a trapezoid is the sum of two triangles.*
> *Factoring out $\frac{1}{2}h$, we get the formula for the area of the trapezoid.*

If the legs are equal, the trapezoid is called an **isosceles trapezoid**, shown below.

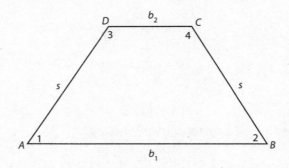

An isosceles trapezoid has the following additional properties:

* Perimeter $= p = b_1 + b_2 + 2s$.

* The diagonals are equal, $AC = BD$.

* The base angles are equal, $\angle 1 = \angle 2$ and $\angle 3 = \angle 4$.

**Example 14:** Find the area and the perimeter of Figure *ABCD*.

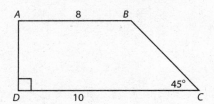

**Solution:**   Draw the height from point *B*, *BG*, as shown below.

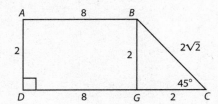

$DG = 8$; $CG = 2$. Because $\triangle BGC$ is an isosceles right triangle, the height

*BG* (and *AD*) $= 2$; *BC*, the hypotenuse of the isosceles right triangle, is

therefore $2\sqrt{2}$. Therefore, the perimeter is $p = 10 + 2 + 8 + 2\sqrt{2} =$

$20 + 2\sqrt{2}$, and the area is $A = \dfrac{1}{2}h(b_1 + b_2) = \dfrac{1}{2}(2)(8 + 10) = 18$.

**Example 15:** Given trapezoid *ORST*, with $RS \parallel OT$, find the coordinates of point *S*. Find the perimeter and the area of trapezoid *ORST*.

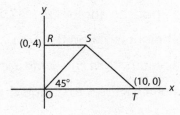

**Solution:**   Because $\triangle ORS$ is a 45°-45°-90° triangle, $OR = RS = 4$, so *S* is the point

$(4, 4)$. The length of $OT = 10$. By the distance formula, the length of

$ST = \sqrt{(10 - 4)^2 + (0 - 4)^2} = \sqrt{52} = 2\sqrt{13}$. Therefore, the perimeter is

$p = 10 + 4 + 4 + 2\sqrt{13}$, or, to be fancy, $2(9 + \sqrt{13})$. Area $= \dfrac{1}{2}h(b_1 + b_2)$

$= \dfrac{1}{2}(4)(10 + 4) = 28$.

**Example 16:** Find the area of isosceles trapezoid *EFGH*.

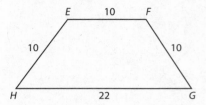

**Solution:** Draw in the two heights for the trapezoid.

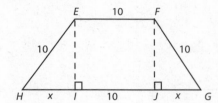

The two bases of the triangles formed are equal because it is an isosceles trapezoid. From the figure, $2x + 10 = 22$, so $x = 6$. Each of the triangles is a 6-8-10 Pythagorean triple, so the height of the trapezoid is 8. Therefore, $A = \frac{1}{2}(8)(10 + 22) = 128$.

# Chapter 11 Quiz

1. In a parallelogram, one angle is $3x - 7$. The opposite angle is $x + 37$. Find the four angles.
2. The bases of a trapezoid are 10 and 8. If the height is 7, find the area.
3. The side of a square is 10. Find the base of a rectangle of the same area if the height is 12.
4. The angles of a quadrilateral are in the ratio of 4 to 5 to 5 to 6. Find the angles.

For questions 5–7, *ABCD* is a quadrilateral with $A(1, 4)$, $B(10, 9)$, $C(14, 15)$, $D(5, 10)$.

5. Show that quadrilateral *ABCD* is a parallelogram from the definition.
6. Show that the opposite sides of this quadrilateral are equal.
7. Show that the diagonals of this parallelogram bisect each other.
8. A parallelogram has base 20, slanted side 10, and the angle between them is 30°. Find its area.
9. The area of a square is 200. Find its diagonal.
10. If you double the base and triple the height, what happens to the area of a rectangle?

11. A rectangle is 5 yards by 2 yards. If it cost $4.00 a square foot to carpet the rectangle, how much will you pay?

12. A rhombus has diagonals 12 and 16. Find its area.

13. The diagonal of a rectangle is 25 and the width is 7. What is the perimeter?

14. The diagonal of a rectangle is 11, and one side is 9. Find its area.

15. For question 14, find the perimeter of the rectangle.

16. The perimeter and area of a square are the same number. What is the length of its side?

17. A trapezoid and a triangle have the same height. One base of the trapezoid equals the base of the triangle. Why is the area of the trapezoid always larger?

18. The perimeter of a rectangle is 24 and the area is 20. Find the length and width.

## Answers to Chapter 11 Test

1. $3x - 7 = x + 37$; $x = 22$; the angles are 59°, 59°, 121°, and 121°.

2. $A = \frac{1}{2}h(b_1 + b_2) = \frac{7}{2}(10 + 8) = 63$.

3. $s^2 = bh$; $10^2 = 12b$; $b = 8.\overline{3}$

4. $4x + 5x + 5x + 6x = 360°$; 72°, 90°, 90°, 108°.

5. Opposite sides must be parallel; the slopes of $AB$ and $CD$ are the same $\left(\frac{5}{9}\right)$, and the slopes of $AD$ and $BC$ are the same $\left(\frac{3}{2}\right)$.

6. $AB = CD = \sqrt{106}$, and $AD = BC = 2\sqrt{13}$.

7. The midpoint of $AC$ and $BD$ is (7.5, 9.5); since they are both the same point, they bisect each other.

8. We form a 30°, 60°, 90° right triangle; the side opposite the 30° angle is half the hypotenuse ( which is 10); so the height is 5, and the area is 100.

9. $s^2 = 200$; $s = 10\sqrt{2}$; so $d = s\sqrt{2} = 10\sqrt{2} \times \sqrt{2} = 20$.

10. Multiplied by 6.

11. 15 by 6, or 90 square feet times $4.00 = $360.

12. $A = \frac{1}{2}d_1 \times d_2 = 96$.

13. By the Pythagorean theorem, the length is $\sqrt{25^2 - 7^2} = \sqrt{625 - 49} = \sqrt{576} = 24$. The perimeter is $(2)(24) + (2)(7) = 62$.

14. Other side $= \sqrt{11^2 - 9^2} = 2\sqrt{10}$. $A = 9 \times 2\sqrt{10} = 18\sqrt{10}$.

15. $P = 18 + 4\sqrt{10}$ or $2(9 + 2\sqrt{10})$.

16. $s^2 = 4s$; $s = 4$.

17. Area of a triangle $= \dfrac{1}{2}bh$.

   Area of a trapezoid $= \dfrac{1}{2}h(b + b_2)$

   $$\dfrac{1}{2}bh + \dfrac{1}{2}b_2h > \dfrac{1}{2}bh.$$

18. $2x + 2y = 24$; $x + y = 12$; $y = 12 - x$; $A = bh = x(12 - x) = 20$; $x^2 - 12x + 20 = (x - 10)(x - 2) = 0$; the dimensions are 10 and 2.

Now, let's look at circles.

**Answer to "Bob Asks":** It is the same: 4% less. To show that this is true, start with $100; an increase of 20% gives $120, and 20% of this is $24. So the net is $96. Now start with $100 and decrease it by 20%, which is $80; 20% of $80 is $16, so we get $80 + $16 = $96.

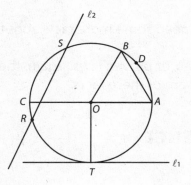

*"An octagon has a perimeter of 5 feet. If 3 inches are added to each side, what is the perimeter of the new octagon?"*

## Circles allow for many short questions that can be combined with the other geometric shapes. Let's get started.

## PARTS OF A CIRCLE

We all know what a circle looks like, but maybe we're not familiar with its "parts."

*O* is the **center** of the circle. A circle is often named by its center, so this is circle *O*.

*OA, OT, OC,* and *OB* are **radii** (singular: radius); a radius is a line segment from the center to the **circumference**, or edge, of the circle.

 *All radii (r) of a circle are equal. This is a postulate or axiom, a law taken to be true without proof.*

It is a good idea to tell you there are no proofs on the COMPASS, as there probably were when you took geometry.

*AC* is the **diameter**, *d*, the distance from one side of the circle through the center to the other side; $d = 2r$ and $r = \dfrac{d}{2}$.

$\ell_1$ is a **tangent**, a line that touches a circle in one and only one point, *T* here.

*T* is a **point of tangency**, the point where a tangent touches the circle. The radius to the point of tangency (*OT*) is always perpendicular to the tangent, so *OT* ⊠ $\ell_1$.

$\ell_2$ is a **secant**, a line that passes through a circle in two places.

*RS* is a **chord**, a line segment that has each end on the circumference of the circle. The diameter is the longest chord in a circle.

*OADBO* is a **sector** (a pie-shaped part of a circle). There are a number of sectors in this figure; others include *BOCSB* and *OATRCSBO*. We will see these again soon.

An **arc** is any distance along the circumference of a circle.

>  Arc *ADB* is a **minor arc** because it is less than half a circle. Arc *BDATRC* is a **major arc** because it is more than half a circle.
>
>  Arc *ATRC* is a **semicircle** because it is exactly half a circle. Arc *ADBSC* is also a semicircle.

Whew! Enough! However, we do need some more facts about circles.

The following are mostly theorems, or proven laws. Again, there are no proofs on the COMPASS, but you need to be aware of these facts.

## AREA AND CIRCUMFERENCE

Area of a circle:

$$A = \pi r^2$$

Circumference (perimeter of a circle):

$$C = 2\pi r \text{ or } \pi d$$

## SECTORS

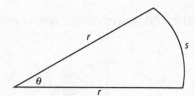

Area of a sector:

$$A = \frac{\theta}{360°}\pi r^2,$$

where θ (theta) is the angle of the sector in degrees.

Arc length of a sector:

$$s = \frac{\theta}{360°} \times 2\pi r$$

Perimeter of a sector:

$$p = s + 2r,$$

where *s* is the arc length.

**Example 1:**   Find the area and perimeter of a 60° sector of a circle of diameter 12.

**Solution:**

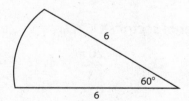

If the diameter is 12, the radius is 6. The sector is pictured above. Its area is $A = \dfrac{60°}{360°} \times \pi 6^2 = 6\pi$ square units. Although you should know that pi (π) is about 3.14, I've never seen a problem for which you had to multiply 6 times 3.14. The answer is left in terms of π. The perimeter of the sector is $s = 2(6) + \dfrac{60°}{360°} \times 2\pi(6) = 12 + 2\pi$ units.

**Q**  **Let's do some multiple-choice exercises.**

For Exercises 1 through 5, refer to the following circle, with center O.

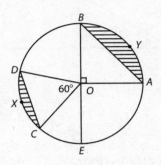

**Exercise 1:**  If the area of △AOB is 25, the area of circle O is

A. $12.5\pi$            D. $100\pi$

B. $25\pi$              E. $200\pi$

C. $50\pi$

**Exercise 2:**  If OA = 8, the area of the shaded region BAYB is

A. $16(\pi - 2)$        D. $64(\pi - 1)$

B. $16(\pi - 1)$        E. $32(\pi - 2)$

C. $8(2\pi - 1)$

**Exercise 3:**  If CD = 10, the perimeter of sector DOCXD is

A. $30 + \dfrac{10\pi}{3}$        D. $20 + \dfrac{20\pi}{3}$

B. $30 + \dfrac{20\pi}{3}$        E. $30 + 30\pi$

C. $20 + \dfrac{10\pi}{3}$

**Exercise 4:**  If OC = 2, the area of the shaded portion DCXD is

A. $\pi - \sqrt{3}$         D. $\dfrac{2\pi - 3\sqrt{3}}{3}$

B. $2\pi - \sqrt{3}$        E. $\dfrac{8\pi - 3\sqrt{3}}{3}$

C. $4\pi - \sqrt{3}$

**Exercise 5:**   If the area of $\triangle COD$ is $25\sqrt{3}$, the perimeter of semicircle $EOBDXCE$ is

A. $10\pi$                          D. $10(2\pi + 1)$

B. $10(\pi + 1)$                    E. $20(\pi + 1)$

C. $10(\pi + 2)$

I guess you get the idea already.

 **Let's look at the answers.**

**Answer 1:**   C: $A = \dfrac{1}{2}(r)(r) = \dfrac{1}{2}r^2 = 25$, so $r^2 = 50$. The area of the circle is thus $\pi r^2 = 50\pi$. Notice that once we have a value for $r^2$, we don't have to find $r$ to do this problem.

**Answer 2:**   A: The area of region $BAYB$ is the area of one-fourth of a circle minus the area of $\triangle AOB$. So the area is $A = \dfrac{1}{4}\pi 8^2 - \dfrac{1}{2}(8)(8) = 16\pi - 32 = 16(\pi - 2)$.

**Answer 3:**   C: The perimeter of sector $DOCXD = 2r + s$, where $s$ is the length of arc $CXD$. $\triangle COD$ is equilateral, so $CD = CO = DO = r = 10$. $s = \dfrac{60°}{360°}2\pi(10) = \dfrac{10\pi}{3}$. So the perimeter of sector $DOCXD$ is $p = 20 + \dfrac{10\pi}{3}$.

**Answer 4:**   D: The area of region $DCXD$ is the area of sector $ODXCO$ minus the area of $\triangle DOC$, when $OC = 2$. So the area is

$$A = \dfrac{60°}{360°}\pi 2^2 - \dfrac{2^2\sqrt{3}}{4} = \dfrac{2\pi}{3} - \sqrt{3} = \dfrac{2\pi - 3\sqrt{3}}{3}.$$

**Answer 5:**   C: We must first find the radius. The area of equilateral $\triangle COD = \dfrac{s^2\sqrt{3}}{4} = 25\sqrt{3}$. So $s^2 = 100$, and $s = r = 10$. The perimeter of the semicircle is $\dfrac{1}{2}(2\pi r) + 2r = \pi r + 2r = 10\pi + 20 = 10(\pi + 2)$.

There are a few more things we need to know. When we talked about two intersecting line segments earlier, we saw that, for the figure below, *CE* might equal *ED*; however, if the description of the figure doesn't say so, you cannot assume it. Also, *ED* might be perpendicular to *AB*, but if it doesn't say so, you cannot assume it, either. In fact, we can say *CD* **bisects** *AB* at *E* only if we know that *AE* = *EB* or *E* is the **midpoint** of *AB*.

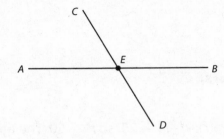

Now, however, we consider two intersecting line segments in a circle, such as chord *AB* and radius *CO* in circle *O*.

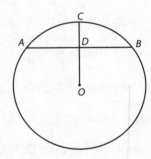

If one of the following facts is true, all are true:

1.  *OD* ⊥ *AB*

2.  *CO* bisects *AB* at point D.

3.  *CO* bisects $\overset{\frown}{ACB}$ (read "arc ACB")

 *When the COMPASS says "distance to a chord," distance always means **perpendicular** distance. Sometimes the COMPASS uses the right-angle sign to show lines are perpendicular (see the figure for Exercise 6). When it doesn't, you have to know it is perpendicular from the information given.*

**Let's do some more multiple-choice exercises.**

For Exercises 6 and 7, use this figure, which is a triangle-semicircle shape.

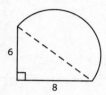

**Exercise 6:**   The perimeter of this figure is

     **A.** $14 + 5\pi$          **D.** $24 + 10\pi$

     **B.** $14 + 10\pi$        **E.** $12 + 10\pi$

     **C.** $24 + 5\pi$

**Exercise 7:**   The area of this figure is

     **A.** $24 + \dfrac{25\pi}{2}$       **D.** $48 + 25\pi$

     **B.** $24 + 25\pi$        **E.** $48 + 50\pi$

     **C.** $24 + 50\pi$

**Exercise 8:**   A circle is **inscribed** in (inside and touching) figure *MNPQ*, which has all right angles. Diameter $AB = 10$. The area of the shaded portion is

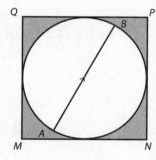

     **A.** $100 - 12.5\pi$      **D.** $40 - 5\pi$

     **B.** $100 - 25\pi$       **E.** $100 - 100\pi$

     **C.** $40 - 10\pi$

**Exercise 9:**   If $AB = 10$, the area of the shaded portion in the figure is

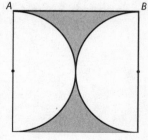

     **A.** $100 - 12.5\pi$      **D.** $40 - 5\pi$

     **B.** $100 - 25\pi$       **E.** $100 - 100\pi$

     **C.** $40 - 10\pi$

For Exercises 10 and 11, use this figure. The perimeter of the 16 semicircles is $32\pi$.

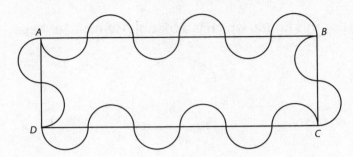

**Exercise 10:** The area of rectangle *ABCD* is

    **A.** 64             **D.** $16\pi$

    **B.** 128          **E.** Cannot be determined

    **C.** 192

**Exercise 11:** The area inside the region formed by the semicircular curves from *A* to *B* to *C* to *D* and back to *A* is

    **A.** 64             **D.** $16\pi$

    **B.** 128          **E.** Are you for real??!!

    **C.** 192

**Exercise 12:** In the figure, $EF = CD = 12$, *B* is the midpoint of *OD*, and *A* is the midpoint of *CO*. The area of the shaded portion is

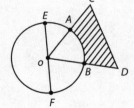

    **A.** $36(4\sqrt{3} - \pi)$        **D.** $72(3\sqrt{3} - \pi)$

    **B.** $6(6\sqrt{3} - \pi)$         **E.** $36(6\sqrt{3} - \pi)$

    **C.** $144(\pi - 3)$

For Exercises 13 and 14, use this figure.

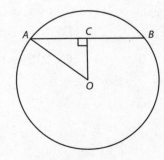

**Exercise 13:**  If the diameter = 20, $OC \perp AB$, and $\angle A = 30°$, $AB =$

  **A.** 10          **D.** $10\sqrt{3}$

  **B.** 5          **E.** $5\sqrt{2}$

  **C.** $5\sqrt{3}$

**Exercise 14:**  If $OC$ bisects $AB$, $AB = 16$, and $OC = 6$, the area of circle $O$ is

  **A.** $10\pi$          **D.** $100\pi$

  **B.** $20\pi$          **E.** $400\pi$

  **C.** $40\pi$

For Exercises 15 and 16, use the following figure of a track.

This track consists of two congruent parallel sides and two equivalent semicircles. The length of each side is 150 meters and the distance between the sides is 40 meters.

  **Exercise 15:**  What is the track's perimeter, in meters?

    **A.** $300 + 40\pi$          **D.** $380 + 20\pi$

    **B.** $380 + 160\pi$          **E.** $300 + 160\pi$

    **C.** $380 + 40\pi$

  **Exercise 16:**  What is the track's area, in square meters?

    **A.** $6,000 + 1,600\pi$          **D.** $3,000 + 1,600\pi$

    **B.** $6,000 + 800\pi$          **E.** $3,000 + 800\pi$

    **C.** $6,000 + 400\pi$

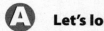 **Let's look at the answers.**

**Answer 6:**  **A:** The figure includes a 6-8-10 Pythagorean triple, but $d = 10$ is not part of the perimeter. However, we need the radius $r = \frac{1}{2}d = 5$ for the circle. So $p = 6 + 8 + \frac{1}{2}2\pi(5) = 14 + 5\pi$.

**Answer 7:**    A: $A = \dfrac{1}{2}bh + \dfrac{1}{2}\pi r^2 = \dfrac{1}{2}(6\times 8)\,\dfrac{1}{2}\pi 5^2 = 24 + \dfrac{25\pi}{2}$.

**Answer 8:**    B: The shaded area is the area of the square minus the area of the circle. $A = s^2 - \pi r^2 = 10^2 - \pi 5^2 = 100 - 25\pi$.

**Answer 9:**    B: Exercises 8 and 9 are exactly the same problems. In Exercise 8, we could also say the square **circumscribes** the circle.

**Answer 10:**    C: Each semicircle has arc length $\dfrac{180°}{360°}\pi d$, and there are 16 semicircles, so $16\left(\dfrac{1}{2}\pi d\right) = 32\pi$, or $8\pi d = 32\pi$, so $d = 4$. The rectangle's dimensions are thus 8 and 24. The area is $A = b \times h = 8(24) = 192$.

**Answer 11:**    C: Believe it or not, Exercise 11 is exactly the same as Exercise 10! We can think of the areas of the "outer" semicircles as canceling out the areas of the "inner" semicircles, and we are left with only the area of rectangle *ABCD*.

**Answer 12:**    B: The information is enough to tell us the triangle is equilateral and $\angle AOB = 60°$. The shaded area is the area of $\triangle COD$ minus the area of sector *OABO*.

Thus, $A = \dfrac{s^2\sqrt{3}}{4} - \dfrac{1}{6}\pi r^2 = \dfrac{12^2\sqrt{3}}{4} - \dfrac{1}{6}\pi 6^2 = 36\sqrt{3} - 6\pi = 6(6\sqrt{3} - \pi)$.

**Answer 13:**    D: *AO*, the radius, is 10; *CO*, the side opposite the 30° angle, is 5; and *AC*, the side opposite the 60° angle, is $5\sqrt{3}$. $AB = 2(AC) = 2(5\sqrt{3}) = 10\sqrt{3}$.

**Answer 14:**    D: To find the area of the circle, we need to find the radius *OA*. We know *AC* is 8 and *OC* is 6. We have a 6-8-10 right triangle, so $AO = r = 10$. The area is $\pi(10)^2 = 100\pi$.

**Answer 15:**    A: The distance between the parallel sides is the diameter of each semicircle, which is 40 meters. The combined length of the two semicircles is equivalent to the length of one complete circle with a diameter of 40 meters. The perimeter of the track is then the sum of the two parallel sides plus the circumference of a circle with a diameter of 40 meters. Thus, the perimeter is $(2)(150) + (\pi)(40) = 300 + 40\pi$ meters.

**Answer 16:**    C: Note that the radius of a circle with a diameter of 40 meters is $(\dfrac{1}{2})(40)$ = 20 meters. The area of the track is the area of a rectangle with length and width of 150 and 40 respectively, plus the area of a circle with a radius of 20 meters. Thus, the area is $(150)(40) + (\pi)(20^2) = 6{,}000 + 400\pi$ square meters.

## Chapter 12 Quiz

All answers may be left in terms of $\pi$.

1. A line segment from one side of the circle to the other is called what?

2. A line hitting a circle in two places is called what?

3. What is the part of the circle that is larger than a semicircle?

4. The point at which the tangent line touches the circle is called what?

5. A line segment from the center of a circle to the circle is perpendicular to a chord. What other two facts are true about this line segment?

6. Find the area of a circle if the diameter is 14.

7. Find the circumference of a circle if the radius is 7.6.

8. What is the diameter of the circle if the area and circumference are the same number?

Questions 9–11 refer to a circle with a 20° sector and a diameter of 20.

9. Find the area of the sector.

10. Find the length of the arc.

11. Find the perimeter of the sector.

Questions 12 and 13 refer to a Norman window (a rectangle with a semicircle on top). The semicircle is on top of the side of the rectangle that is $4x$; the other side of the rectangle is $8x$.

12. Find the perimeter of the Norman window in terms of $x$ and $\pi$.

13. Find the area of the Norman window in terms of $x$ and $\pi$.

Questions 14 and 15 refer to two concentric circles (having the same center).

14. If the inner diameter is 14 and the outer diameter is 16, find the area between the circles.

15. If the inner radius is 10 and the width between the circles is 2, find the area between the circles.

16. A circle is inside a square, touching the square in 4 points. The side of the square is 6. Find the area inside the square but outside the circle.

17. A square is inside a circle, touching the circle in 4 points. The side of the square is 6. Find the area inside the circle and outside the square.

18. A circle has area $20\pi$. If a sector has area $4\pi$, how many degrees are in the central angle of the sector?

19. A chord of length 16 is perpendicular to a line segment whose endpoints are the center of the circle and a point on the chord. The segment has length 6. Find the area of the circle.

### Answers to Chapter 12 Quiz

1. Chord.

2. Secant.

3. Major arc.

4. Point of tangency.

5. It bisects the chord and it bisects the arc.

6. $49\pi$.

7. $15.2\pi$.

8. $\pi r^2 = 2\pi r$; $r = 2$; so $d = 4$.

9. $\dfrac{20}{360} \times \pi \times 10^2 = \dfrac{50}{9}\pi$.

10. $\dfrac{20}{360} \times 2\pi \times 10 = \dfrac{10\pi}{9}$.

11. $20 + \dfrac{10\pi}{9}$.

12. $8x + 8x + 4x + \left(\dfrac{1}{2}\right)2\pi(2x) = 20x + 2\pi x$.

13. $(4x)(8x) + \left(\dfrac{1}{2}\right)\pi(2x)^2 = 32x^2 + 2\pi x^2$.

14. $\pi\left(r^2_{outside} - r^2_{inside}\right) = \pi(8^2 - 7^2) = 15\pi$.

15. $\pi(12^2 - 10^2) = 44\pi$.

16. $r = 3$; $A = 36 - 9\pi$.

17. $r = \dfrac{6}{\sqrt{2}} = 3\sqrt{2}$; $A = 18\pi - 36$.

18. $\dfrac{4\pi}{20\pi} = \dfrac{1}{5}$ the circle $= 72°$.

19. The segment bisects the chord (8). The picture is a 6–8–10 right triangle, where the radius is 10. The area of the circle is $100\pi$.

Okay. Now let's go from two dimensions to three dimensions.

**Answer to "Bob Asks":** 7 feet. It is not necessary to know how long each side is. Adding 3 inches to each of 8 sides means 24 inches, or 2 feet, are added to the original perimeter.

# CHAPTER 13: *All About Three-Dimensional Figures*

**"**A square has a perimeter of 1 foot. If we have a second square with each side half as large, what is the perimeter of the smaller square? **"**

**This** chapter is relatively short. There are only a few figures we need to know. Because these are three-dimensional figures, we discuss their volumes and surface areas (areas of all of the surfaces). The diagonal is the distance from one corner internally to an opposite corner.

## BOX

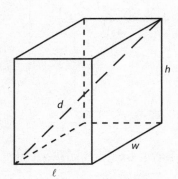

This figure is also known as a **rectangular solid**, and if that isn't a mouthful enough, its correct name is a **rectangular parallelepiped**. But essentially, it's a **box**.

- Volume = $V = \ell wh$
- Surface area = $SA = 2\ell w + 2\ell h + 2wh$
- Diagonal = $d = \sqrt{\ell^2 + w^2 + h^2}$, known as the 3-D Pythagorean Theorem

**Example 1:**    For the given figure, find V, SA, and d.

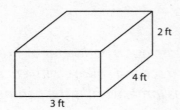

**Solution:**    $V = \ell wh = (3)(4)(2) = 24$ cubic feet; $SA = 2\ell w + 2\ell h + 2wh = 2(3)(4) +$
$2(3)(2) + 2(4)(2) = 52$ square feet; $d = \sqrt{\ell^2 + w^2 + h^2} = \sqrt{3^2 + 4^2 + 2^2}$
$= \sqrt{29} \approx 5.4$ feet

# CUBE

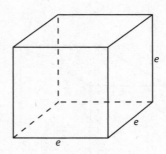

A **cube** is a box for which all of the faces, or sides, are equal squares.

- $V = e^3$ (Read as "e cubed.") Cubing comes from a cube!
- $SA = 6e^2$
- $d = e\sqrt{3}$
- A cube has 6 faces, 8 vertices, and 12 edges.

**Example 2:**    For a cube with an edge of 10 meters, find V, SA, and d.

**Solution:**    $V = 10^3 = 1,000$ cubic meters; $SA = 6e^2 = 6(10)^2 = 600$ square meters;
$d = e\sqrt{3} = 10\sqrt{3} \approx 17.32$ meters

## CYLINDER

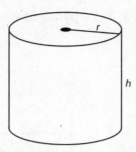

A cylinder is shaped like a can. The curved surface is considered as a side, and the top and bottom are equal circles.

- $V = \pi r^2 h$

- $SA = \text{top} + \text{bottom} + \text{curved surface} = 2\pi r^2 + 2\pi rh$

Once a neighbor of mine wanted to find the area of the curved part of a cylinder. He wasn't interested in why, just the answer. Of course, being a teacher, I had to explain it to him. I told him that if he cut a label off a soup can and unwrapped it, the figure is a rectangle; neglecting the rim, the height is the height of the can and the width is the circumference of the circle. Multiply this height and width, and the answer is $2\pi r \times h$. He waited patiently and then soon moved to another town. (Just kidding!)

In general, the volume of any figure for which the top is the same as the bottom is $V = Bh$, where $B$ is the area of the base. If the figure comes to a point, the volume is $\left(\dfrac{1}{3}\right)Bh$. The surface area is found by adding up all the surfaces.

**Example 3:** Find $V$ and $SA$ for a cylinder of height 10 yards and diameter of 8 yards.

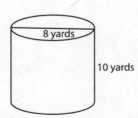

8 yards

10 yards

**Solution:** We see that, because $d = 8$, $r = 4$. Then $V = \pi r^2 h = \pi(4^2 \times 10) = 160\pi$ cubic yards; $SA = 2\pi r^2 + 2\pi rh = 2\pi 4^2 + 2\pi(4)(10) = 112\pi$ square yards.

**ⓠ**    **Let's do some multiple-choice exercises.**

Use this figure for Exercises 1 through 3. It is a pyramid with a square base. $WX = 8$, $BV = 3$, and $B$ is in the middle of the base.

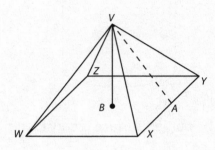

**Exercise 1:**    The volume of the pyramid is

A. 192                      D. 32

B. 96                       E. 16

C. 64

**Exercise 2:**    The surface area of the pyramid is

A. 72                       D. 224

B. 112                      E. 448

C. 144

**Exercise 3:**    $VY =$

A. 6                        D. 9

B. $\sqrt{41}$              E. $\sqrt{89}$

C. 7

**Exercise 4:**    In the given rectangular solid, the perimeter of $\triangle ABC =$

A. $\sqrt{325} = 5\sqrt{13}$    D. 37

B. 30                       E. 41

C. $27 + \sqrt{261}$

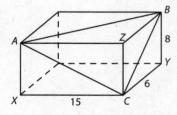

**Exercise 5:** The volume of the cylinder shown is:

A. $640\pi$          D. $144\pi$

B. $320\pi$          E. $72\pi$

C. $288\pi$

**Exercise 6:** *ABKL* is the face of a cube with $AB = 10$, and box *BCFG* has a square front with $BC = 6$. The surface area that can be viewed in this configuration is

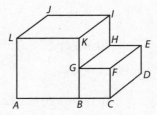

A. 300          D. 400

B. 356          E. 1360

C. 396

**Exercise 7:** A cylinder has volume *V*. If we triple its radius, what do we multiply the height by in order for the volume to stay the same?

A. $\dfrac{1}{9}$          D. 3

B. $\dfrac{1}{3}$          E. 9

C. 1

 **Let's look at the answers.**

**Answer 1:** C: $V = \left(\dfrac{1}{3}\right)Bh = \dfrac{1}{3}s^2h = \dfrac{1}{3}(8^2)(3) = 64.$

**Answer 2:** C: $SA = s^2 + 4\left(\dfrac{1}{2}bh\right).$ $AB = \left(\dfrac{1}{2}\right)WX = 4;$ $\triangle ABV$ is a 3-4-5 right triangle with $AB = 4$ and $BV = 3$, so $AV = 5$. $AV$ is the height of each triangular side, $h$, and $b = XY = 8$. So $SA = 8^2 + 2(8)(5) = 144.$

**Answer 3:** B: $\triangle AVY$ is a right triangle with right angle at $A$. $AY = 4$ and $AV = 5$, so $VY = \sqrt{4^2 + 5^2} = \sqrt{41}.$

**Answer 4:**   **C:** In the given figure, we have to use the 2-D Pythagorean Theorem three times to find the sides of $\triangle ABC$. $\triangle BCY$ is a 6-8-10 triple, so $BC = 10$, and $\triangle ACX$ is a 8-15-17 triple, so $AC = 17$. For $\triangle ABZ$, we actually have to calculate the missing side $AB = \sqrt{6^2 + 15^2} = \sqrt{261}$. So the perimeter is $10 + 17 + \sqrt{261} = 27 + \sqrt{261}$.

**Answer 5:**   **E:** The diameter of the base is 6, and again we have a Pythagorean 6-8-10 triple; so $h = 8$. The volume is $\pi(3^2)(8) = 72\pi$.

**Answer 6:**   **C:** The areas are: $ABKL = 100$; $IJLK = 100$; $BCFG = 36$; $CDEF = 60$; $EFGH = 60$; $GHIK = 40$. The total is 396.

**Answer 7:**   **A:** The volume $V = \pi r^2 h$. For simplicity, let $r = 1$ and $h = 1$. So $V = \pi$.

If we triple the radius, $V = \pi(3)^2 h$. For the original volume to still be $\pi$, $9h = 1$, or $h = \dfrac{1}{9}$.

# SPHERE

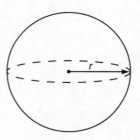

Another figure we need to know the formulas for is the **sphere,** on which all points are the same distance from the center. That distance is called the **radius** $r$ of the sphere.

- $V = \dfrac{4}{3}\pi r^3$

- $SA = 4\pi r^2$

**Example 4:**   If a sphere has radius of 10 inches, find its volume and surface area.

**Solution:**   $V = \dfrac{4}{3}\pi r^3 = \dfrac{4000\pi}{3}$ cubic inches. $SA = 4\pi r^2 = 400\pi$ square inches.

## CONE

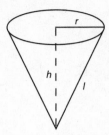

The last 3-D shape you need to know is the **cone;** technically, this is a right circular cone. The word "right" here means the height is to the middle of the base, which is a circle. The **slant height,** which is the distance from any point on the circle to the point at the top of the cone, is designated by $\ell$.

- $V = \dfrac{1}{3}\pi r^2 h$

- $SA = \pi r^2 + \pi r \ell$

- $\ell = \sqrt{r^2 + h^2}$

**Example 5:** Find the volume and surface area of a cone with $r = 8$ and $h = 7$.

**Solution:** We need to find the slant height first. The slant height is

$$\ell = \sqrt{8^2 + 7^2} = \sqrt{113}.\; V = \frac{1}{3}\pi(8^2)(7) = \frac{448\pi}{3} \text{ and}$$

$$SA = \pi 8^2 + \pi(8)\sqrt{113} = 8\pi(8 + \sqrt{113}).$$

 *Every formula involving $\pi$ has been derived with calculus. One thing that thrills many students is to be actually shown that the area of a circle is $\pi r^2$! It really is true!*

## Chapter 13 Quiz

Answers may be left in terms of $\pi$.

Questions 1 and 2 refer to a sphere of diameter 18.

1. Find the volume of the sphere.

2. Find the surface area of the sphere.

Questions 3–5 refer to a box with width 5, length 6, and height 7.

3. Find the volume of the box.

4. Find the surface area of the box.

5. Find the diagonal of the box.

Questions 6 and 7 refer to a cylinder of diameter 10 and height 20.

6. Find the volume of the cylinder.

7. Find the surface area of the cylinder.

Questions 8–10 refer to a cube with edge 12.

8. Find the volume of the cube.

9. Find the surface area of the cube.

10. Find the diagonal of the cube.

Questions 11–13 refer to a cone of diameter 6 and height 4.

11. Find the slant height of the cone.

12. Find the volume of the cone.

13. Find the surface area of the cone.

14. If the volume of a cube is numerically equal to its surface area, what is the length of a side?

15. For what radius is the volume of a sphere equal numerically to its surface area?

Questions 16 and 17 refer to a child's toy block that is a cylinder with a hemisphere (half a sphere) on top of it.

16. Find its surface area if the radius = 8 and the height = 10.

17. Find its volume if the radius = 8 and the height = 10.

18. A pyramid has a rectangular base that is 5 feet by 3 feet with a height of 7 feet. Find its volume.

19. For a cube, verify that the number of sides plus the number of vertices equals the number of edges plus 2.

20. What is the minimum number of colors you need to color a cube so that the same colors don't touch?

### Answers to Chapter 13 Quiz

1. $r = 9; V = \frac{4}{3}\pi \times 9^3 = 972\pi.$

2. $S = 4\pi \times 9^2 = 324\pi.$

3. $V = (5)(6)(7) = 210.$

4. $S = 2[(6)(7) + (5)(7) + (5)(6)] = 214.$

5. Diagonal $= \sqrt{5^2 + 6^2 + 7^2} = \sqrt{110}.$

6. $V = \pi(5)^2 20 = 500\pi.$

7. $S = 2\pi(5)^2 + 2\pi(5)(20) = 250\pi$.

8. 1728.

9. 864.

10. $12\sqrt{3}$.

11. The slant height is 5 because $r = 3$, and it is the hypotenuse of a 3-4-5 right triangle.

12. $V = \frac{1}{3}\pi(3)^2 4 = 12\pi$.

13. $SA = \pi(3)^2 + \pi(3)(5) = 24\pi$.

14. $e^3 = 6e^2$. Thus $e = 6$.

15. $\frac{4}{3}\pi r^3 = 4\pi r^2$. Then $\frac{4}{3}r = 4$. Thus, $r = 3$.

16. $SA = \pi r^2 + 2\pi(r)(h) + \left(\frac{1}{2}\right)4\pi r^2 = (64 + 160 + 128)\pi = 352\pi$.

17. $V = \pi r^2 h + \frac{1}{2} \times \frac{4}{3}\pi r^3 = \pi\left(640 + \frac{1024}{3}\right) = \left(981\frac{1}{3}\right)\pi$.

18. $\frac{1}{3}(5)(3)(7) = 35$.

19. $6 + 8 = 12 + 2$.

20. Three! The top and bottom are one color; the front and the back are another color; and the two sides are a third color. This one is for fun. Yes, in math some questions are fun questions, at least for me and hopefully for you.

Enough for geometry! Let's move on to other topics you might see on the COMPASS.

**Answer to "Bob Asks":** $\frac{1}{2}$; half the sides means half the perimeter. Oh, let's show it. The perimeter of the first square is 1 foot, which means the sides are 3 inches each. Half of that is $1\frac{1}{2}$ inches; times 4 is 6 inches, or $\frac{1}{2}$ foot.

# CHAPTER 14: *More Algebraic Topics*

"*A monkey is trying to climb out of a 50-foot hole. In the morning, it climbs 5 feet. At night, it gets so tired that it falls back 4 feet. The monkey does this every day. On what day does the monkey get out of the hole?*"

## FUNCTIONS

A most important part of algebra is the study of functions. Let's give it its due.

**Function:** To each element in set $D$, we assign one and only one element. The set $D$ is called the **domain**. On the COMPASS, you should think of the $x$ values. The assignment is called the **map**; the set of numbers that are assigned is called the **range**. On the COMPASS, you should think of the $y$ values.

**Note** *If you are unfamiliar with sets and elements, see Chapter 16.*

**Example 1:**  Let $f(x) = x^2 + 3x + 22$; find the following:

| Problem | Solution |
|---------|----------|
| **a.** $f(4)$ | $4^2 + 3(4) + 22 = 50$ |
| **b.** $f(-3)$ | $(-3)^2 + 3(-3) + 22 = 22$ |
| **c.** $f(0)$ | $0^2 + 3(0) + 22 = 22$ |
| **d.** $f(x + h)$ | $(x + h)^2 + 3(x + h) + 22 =$ $x^2 + 2xh + h^2 + 3x + 3h + 22$ |
| **e.** $f(\text{pigs})$ | $(\text{pigs})^2 + 3\text{pigs} + 22$ |

**Example 2:**  Let $g(x) = x^2 + 7x$; let $D = \{-3, 4, 7\}$. Find the range.

**Solution:**  Range $= \{g(-3), g(4), g(7)\} = \{-12, 44, 98\}$

**Note** *Functions (the maps) are usually indicated by f(x), g(x), F(x), and G(x). However, any letter can be used.*

> **Note**  *A graph is a function if any time you draw a vertical line, the line crosses the graph only once. All lines (except a vertical line) are functions; a circle is not a function.*

## PIECEWISE FUNCTIONS

Sometimes a function has more than one piece. Let's give an example; we'll do it the longer way for understanding and then the shorter way.

**Example 3:** Graph $f(x) = \begin{cases} x+6 & \text{if} \quad x < 0 \\ x^2 & \text{if} \quad 0 \le x \le 3 \\ 5-x & \text{if} \quad x > 3 \end{cases}$

**Solution:** We make a table of values: $x = -3, -2, -1, 0^-, 0, 1, 2, 3, 3^+, 4, 5, \ldots$

When $x < 0$, we use the first part of the definition, $f(-3) = (-3) + 6 = 3$, and we have the point $(-3, 3)$.

Similarly, $f(-2) = (-2) + 6 = 4$, and we have the point $(-2, 4)$; $f(-1) = (-1) + 6 = 5$, which yields the point $(-1, 5)$; $f(0^-) = 0^- + 6 = 6^-$ (or a little less than 0, like $-.0001$, plus 6 equals a little less than 6). The point is $(0^-, 6^-)$—on a graph this is an open dot because the graph doesn't quite hit $(0, 6)$.

For the second part of the definition, $f(0) = 0^2 = 0$ yields the point $(0, 0)$; $f(1) = 1^2 = 1$ gives $(1, 1)$; $f(2) = 2^2 = 4$ gives $(2, 4)$; and $f(3) = 9$ gives $(3, 9)$.

For the third part of the definition, $f(3^+) = 5 - 3^+ = 2^-$ (5 minus a little more than 3 equals a little less than 2, or $2^-$), which yields the point $(3^+, 2^-)$ with an open dot on $(3, 2)$; $f(4) = 5 - 4 = 1$ gives $(4, 1)$; and $f(5) = 5 - 5 = 0$ gives $(5, 0)$.

The graph would look like this:

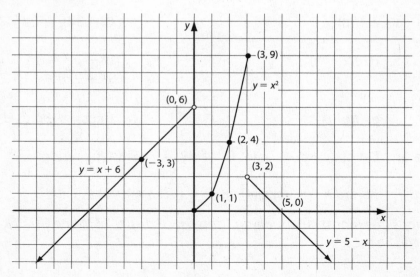

To be able to do this same problem the short way, you must know or be able to get the graph really quickly.

We graph $y = x + 6$, but we are interested only in the part of the line where $x < 0$.

We graph $y = x^2$, but we are interested only in the part of the graph between 0 and 3, inclusive.

We graph $y = 5 - x$, but we are interested only in the part of the graph where $x > 3$.

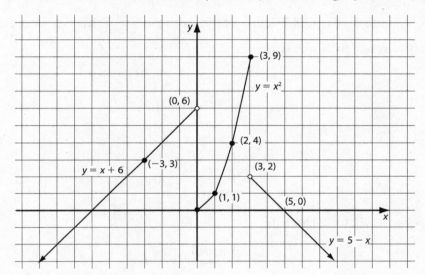

The COMPASS could provide a value for $f(x)$ and you would have to know what part of the definition of a piecewise function holds and find $x$.

The COMPASS will have a question with five answer choices, one of which tells what $x$ is.

 **Let's look at a multiple-choice exercise.**

**Exercise 1:**  Referring to the figure above, if $f(x) = 4$, $x$ could be

A. 2 only

B. −2 only

C. 2, −2 only

D. 1 and 2 only

E. 1, 2, and −2 only

**Let's look at the answer.**

**Answer 1:**   **C:** If $f(x) = x + 6 = 4$; $x = -2$; because $-2 < 0$, $x = -2$ is okay. If $f(x) = x^2 = 4$; $x = 2$ and $x = -2$; only 2 is between 0 and 3; so $x = 2$ is okay. If $f(x) = 5 - x = 4$, $x = 1$; because 1 is not $> 3$, $x = 1$ is not okay.

## PARABOLAS

The parabola you will see on the COMPASS is of the form $y = ax^2 + bx + c$, where $a$ cannot be 0.

- If $a > 0$, the graph will look like this:

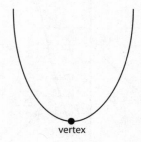

vertex

- If $a < 0$, the graph will look like this:

vertex

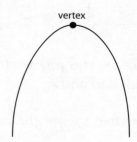

The vertex is the lowest or highest point of the parabola. Its $x$ value is found by setting $x = -\dfrac{b}{2a}$.

**Example 4:**   Sketch $y = x^2 - 4x - 5$.

**Solution:**   We sketch by finding the vertex, $x$-intercept(s) and $y$-intercept.

Vertex: $x = -\dfrac{b}{2a} = -\dfrac{-4}{2} = 2$; $y = (2)^2 - 4(2) - 5 = -9$: $(2, -9)$

$x$-intercept(s): $y = 0$; $x^2 - 4x - 5 = (x - 5)(x + 1) = 0$: $(5, 0)$ and $(-1, 0)$.

$y$-intercept: $x = 0$; $y = -5$: $(0, -5)$.

The sketch looks like this:

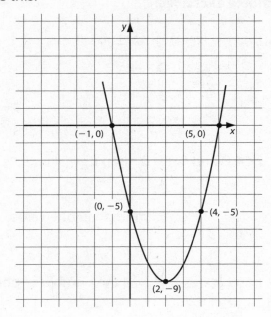

**Example 5:**    Sketch $y = 18 - 2x^2$.

**Solution:**    We sketch by finding the vertex, $x$-intercept(s), and $y$-intercept.

Vertex: $x = -\dfrac{b}{2a} = -\dfrac{0}{2(2)} = 0$; $y = 18$: $(0, 18)$. It is also the $y$-intercept.

$x$-intercepts: $y = 0$; $18 - 2x^2 = -2(x^2 - 9) = -2(x + 3)(x - 3) = 0$:

$(-3, 0)$ and $(3, 0)$.

The sketch looks like this:

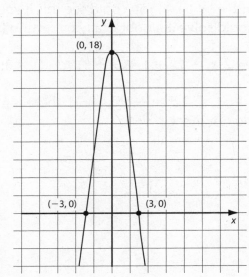

**Note**    *Each box is 2 units.*

**Note**    *With no x term, the parabola is symmetric with respect to the y-axis.*

**Q**    **Let's try a few multiple-choice exercises:**

**Exercise 2:**    $f(x) = (x + 5)^2$, $g(x) = (x + 11)^2$, and $g(x) = f(x)$; $x =$

A. 16                      D. $-8$

B. 8                       E. $-16$

C. 0

**Exercise 3:**    If $f(x) = x^2 + 2x + 3$ and $f(x) = 2$, then $x =$

A. 11                      D. $-1$

B. 1                       E.  more than one value

C. 0

**Exercise 4:**    If $f(x) = x^2 + 1$ and $g(x) = -x^2 - 1$; the number of points where $f(x)$ meets $g(x)$ is

A. 0                       D. 3

B. 1                       E.  more than 3

C. 2

**A**    **Let's look at the answers.**

**Answer 2:**    D: The answer is halfway between the two $x$-intercepts; halfway between $-5$ and $-11$ is $-8$. You could see this if you sketched the curve. Or you could solve it algebraically by setting $(x + 5)^2 = (x + 11)^2$.

**Answer 3:**    D: $x^2 + 2x + 3 = 2$; $x^2 + 2x + 1 = (x + 1)(x + 1) = 0$; so $x = -1$.

**Answer 4:**    A: $x^2 + 1 = -x^2 - 1$; so $2x^2 = -2$; $x^2 = -1$; there is no real solution, so the graphs do not meet.

# ABSOLUTE VALUE GRAPH

The vertex of an absolute value graph occurs when the absolute value = 0.

**Example 6:**   Let's graph $y = |x - 4|$.

**Solution:**   The vertex occurs when $x = 4$, at the point (4, 0). Substituting $x = 2, 3, 5$, and 6, we get the following graph:

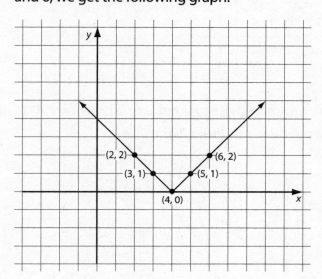

**Example 7:**   Let's graph $y = -|x + 3| + 1$.

**Solution:**   The vertex occurs at $x = -3$; $y = 1$; $(-3, 1)$. Substituting $x = -5, -4, -2$, $-1$, we get the following graph:

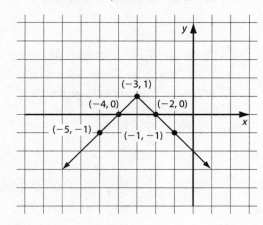

**Example 8:**    Given the graph $y = |x + 2|$ shown here, if we rotated it 90° clockwise, what would the figure look like?

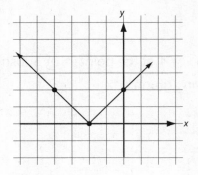

**Solution:**    What you should do is turn the test booklet 90° clockwise to see the answer. The graph looks like this:

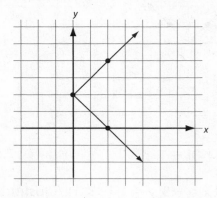

# COMPOSITION

Suppose we have a function with domain $D$, range $R_1$, and map $f$. Also suppose we have a second function with domain $R_1$, range $R_2$, and map $g$. Is there a function that takes us directly from $D$ to $R_2$? The answer is: Of course; otherwise, why would I write this?

The map is called the composition $g \circ f$ (read, "$g$ circle $f$").
The definition is $g \circ f(x) = g(f(x))$.

**Example 9:**    Suppose $f(x) = 5x + 7$ and $g(x) = x^2 + 10$. Find $g(f(4))$, $f(g(4))$, $g(f(x))$, and $f(g(x))$.

**Solution:**    $g(f(4))$: $f(4) = 5(4) + 7 = 27$; $g(27) = 27^2 + 10 = 739$.

$f(g(4))$: $g(4) = 4^2 + 10 = 26$; $f(26) = 5(26) + 7 = 137$.

$f(g(x))$: We know $f(x) = 5x + 7$; so $f(g(x)) = 5[g(x)] + 7 = 5(x^2 + 10) + 7 = 5x^2 + 57$.

$g(f(x))$: We know $g(x) = x^2 + 10$; so $g(f(x)) = [f(x)]^2 + 10 = (5x + 7)^2 + 10 = 25x^2 + 70x + 59$.

**Note** *Maps are not commutative in general. The order in which you do a map matters; you get different answers with different orders.*

**Note** *If we were to draw pictures here, or get technical and indicate domains and ranges, the pictures, domains, and ranges for different orders of the maps would be very different.*

**Note** *If some of you eventually go into math-related topics, if you are doing research in math-related books pre-1970, this definition might be different in those books.*

## INVERSE FUNCTIONS

Suppose we have a function map $f$, domain $D$, and range $R$. Do we have a function that takes $R$ back into $D$. If $D$ and $R$ are in **1–1 correspondence** (can be paired), the answer is yes. It is called the **inverse map** $f^{-1}$ $\left( \text{read, "} f \text{ inverse"; it does } not \text{ mean } \dfrac{1}{f} \right)$.

Graphically, a function (which allows a vertical line to hit the graph only once) can have an inverse only if all horizontal lines can hit the graph no more than once. Let's look at an example.

**Example 10:**    Let $f(x) = 5x + 7$, domain $D = \{1,5,9,20\}$. Find $f^{-1}(x)$, the inverse of $f(x)$.

**Solution:**    First, find the range $R = \{f(1), f(5), f(9), f(20)\} = \{12, 32, 52, 107\}$. Notice that the domain and range are in $1 - 1$ correspondence.

To find the inverse, switch the domain and range. So $D = \{12, 32, 52, 107\}$.

To find the inverse map, we have to solve for $x$. Let $f(x) = y = 5x + 7$. So $5x = y - 7$ and $x = \dfrac{y - 7}{5}$.

Now we have a notation change. The $y$ is in the domain; we rename it $x$. The "$x$" we solved for is the new function $f^{-1}$. So $f^{-1}(x) = \dfrac{x - 7}{5}$ with $D = \{12, 32, 52, 107\}$. Let's check it out.

$$f^{-1}(12) = \frac{12-7}{5} = 1; f^{-1}(32) = \frac{32-7}{5} = 5; f^{-1}(52) = \frac{52-7}{5} = 9;$$
$$\text{and } f^{-1}(107) = \frac{107-7}{5} = 20.$$

 *We have studied inverses in the past. The inverse of adding is subtracting; the inverse of subtracting is adding; the inverse of multiplying is dividing (excluding 0); and the inverse of dividing is multiplying. There are more. If we restrict the domain, the inverse of squaring (or any even root) is square rooting or any even "rooting." For odd powers, no restriction is necessary. The inverse of cubing is cube rooting, etc. Shortly, we will study another inverse. The inverse of exponentiation is logarithms.*

## 1 TO 1 FUNCTION (1–1 FUNCTION)

This topic is related to 1–1 correspondence. We will do this formally.

**1–1 function**: If $f(a) = f(b)$, then $a = b$.

Let's give examples of one function that is 1–1 and one that isn't.

**Example 11:**   Let $f(x) = 2x$. Is $f(x)$ a 1–1 function?

**Solution:**   If $f(a) = f(b)$, in symbols, $2a = 2b$, does $a = b$ always? The answer is yes since we can cancel the 2's. So $f(x) = 2x$ is a 1–1 function.

**Example 12:**   Let $f(x) = x^2$. Is $f(x)$ a 1–1 function?

**Solution:**   If $f(a) = f(b)$, in symbols $a^2 = b^2$, does $a = b$ always? The answer is no! $a = b$ or $a = -b$.

If we draw a picture of Example 11, it is a function that always increases. Curves that always increase or always decrease are 1–1 functions.

For Example 12, the curve both increases and decreases. It is *not* 1–1. By restricting the domain, we can make it 1–1. For example, we can allow only positive values for $x$.

 *We have already used 1–1 without mentioning it. If $2^{3x+7} = 2^{5x-3}$, then $3x + 7 = 5x - 3$. Here $f(x) = 2^x$. If $f(a) = f(b)$, $2^a = 2^b$, and then $a = b$. If you draw $2^x$, it is an increasing function.*

**Example 13:** Let $f(x) = x^2 + 5x + 9$, $g(x) = 4x + 2$, $F(x) = \dfrac{x}{x + 3}$, $G(x) = \dfrac{1}{x}$. Find the values of the following functions:

| Problem | Solution |
|---------|----------|

**a.** $f(4)$      $4^2 + 5(4) + 9 = 45$

**b.** $g\left(-\dfrac{1}{2}\right)$      $4\left(-\dfrac{1}{2}\right) + 2 = 0$

**c.** $f(g(3))$      $f(14) = 275$

**d.** $g(f(1))$      $g(15) = 62$

**e.** $g(f(x))$      $g(x^2 + 5x + 9) = 4(x^2 + 5x + 9) + 2 = 4x^2 + 20x + 38$

**f.** $f(g(x))$      $f(4x + 2) = (4x + 2)^2 + 5(4x + 2) + 9 = 16x^2 + 36x + 23$

**g.** $F(G(x))$      $\dfrac{G(x)}{G(x) + 3} = \dfrac{\frac{1}{x}}{\frac{1}{x} + 3} \times \dfrac{x}{x} = \dfrac{1}{1 + 3x}$

**h.** $F^{-1}(x)$      First, find $F(x) = y = \dfrac{x}{x + 3}$; so $(x + 3)y = xy + 3y = x$;

$xy - x = x(y - 1) = -3y$; so $x = \dfrac{-3y}{y - 1}$ or $\dfrac{3y}{1 - y}$.

We now need a notation change. $F^{-1}(x) = \dfrac{3x}{1 - x}$.

**i.** $\dfrac{f(x + h) - f(x)}{h}$      $\dfrac{(x + h)^2 + 5(x + h) + 9 - (x^2 + 5x + 9)}{h} =$

$\dfrac{2xh + h^2 + 5h}{h} = 2x + h + 5$

The last question is a problem I hope you've seen. It is called the difference quotient. Some students study this in Algebra 2 or Precalc without knowing why. (You should always ask why or why it's true.) Here's the secret. *Shh*. It's at the beginning of calculus. *Shh!*

Let's go on to logarithms.

## LOGARITHMS

Let us first define what is meant by logarithm. We write

$$\log_b x = y$$

(read as log(arithm) of $x$ to the base $b$ is $y$) if $b^y = x$. In words, $\log_{\text{base}}$ answer = exponent, because base$^{\text{exponent}}$ = answer.

 **Note**    *This is a very strange definition and must be practiced, especially if you haven't had much experience with it.*

**Example 14:**   What is the value of $\log_3 81$?

**Solution:**     $\log_3 81 = 4$ because $3^4 = 81$.

The **base** may be only positive numbers other than 1. There are two common bases. If we write "log," it means $\log_{10}$. The reason we use 10 as the base is because we have 10 fingers (really!). If we write "ln," it means $\log e$, where $e$ is a number like pi; $e \approx$ (approximately equals) 2.7. Although it doesn't stand for it, anything to do with $e$ is easy!

The $y$ values, the logarithms, can be any real number.

The $x$ values must be positive because a positive number to any power **must** be positive!

**Example 15:**  Find $x$:

| Problem | Solution |
|---|---|
| **a.** $\log_5 125 = x$ | $5^x = 125$, so $x = 3$ |
| **b.** $\log_4 x = 3$ | $x = 4^3 = 64$ |
| **c.** $\log_x 81 = 2$ | $x^2 = 81$, so $x = 9$ (only, because bases can't be negative) |

Because logarithms are exponents, the laws of logarithms are the laws of exponents.

| Law | Example |
|---|---|

**1.** When you multiply, you add exponents; the same is true for logs.

$$\log_b cd = \log_b c + \log_b d$$

$\log 6 = \log (2)(3) = \log 2 + \log 3;$
$\ln dry = \ln d + \ln r + \ln y$

**2.** When you divide, you subtract exponents; the same is true for logs.

$$\log_b \frac{c}{d} = \log_b c - \log_b d$$

$\log_7 \dfrac{3}{5} = \log_7 3 - \log_7 5$

**3.** A power to a power? You multiply exponents; the same is true for logs.

$$\log_b c^d = d \log_b c$$

$\log 128 = \log 2^7 = 7 \log 2$

**4.** Any number to the first power is that number.

$$\log_b b = 1, \text{ because } b^1 = b.$$

$\log_3 3 = 1,$ because $3^1 = 3;$ $\log 10 = 1$ because $10^1 = 10;$ $\ln e = 1,$ because $e^1 = e.$

**5.** Any nonzero number to the zero power is 1.

$$\log_b 1 = 0, \text{ because } b^0 = 1, \text{ where } b \text{ is any base.}$$

$\log_2 1 = 0,$ because $2^0 = 1;$ $\ln 1 = 0$ because $e^0 = 1$

**6.** Log is a 1–1 function.

If $\log_b x = \log_b 3,$ then $x = 3.$

**7.** Log is an increasing function.

$\log x < \log 6,$ if $x < 6.$

**8.** $b^{\log_b x}$ is a really strange way to write $x$.

$9^{\log_9 5} = 5.$

**9.** $\log_b b^x = x.$

$\log_7 7^5 = 5.$

**10.** $\log_c d = \dfrac{\log_b d}{\log_b c}.$

$\log_3 7 = \dfrac{\log_{10} 7}{\log_{10} 3}.$

The last law needs some explanation. We do not know anything about base 3 (only about 10 and $e$). We can rewrite the problem as log 7 divided by log 3. This is *not, Not, NOT* $\log \left( \dfrac{7}{3} \right) = \log 7 - \log 3.$

In calculus, if you can do the next example and absolutely no other, it is more than 50 percent of what you need to know about logs.

**Example 16:** Write as simpler logs with no (or fewer) exponents: $\log_b \dfrac{p^4 q^9}{v^3 \sqrt{u}}$

**Solution:** We add the logs that are products on the top, subtract the products on the bottom; exponents come down as coefficients:

$$\log_b \frac{p^4 q^9}{v^3 \sqrt{u}} = \log_b(p^4 q^9) - \log_b(v^3 u^{\frac{1}{2}}) = \log_b p^4 + \log_b q^9 - (\log_b v^3 + \log_b u^{\frac{1}{2}}) =$$

$$4\log_b p + 9\log_b q - 3\log_b v - \frac{1}{2}\log_b u.$$

Unfortunately, in Algebra 2 or Precalc, this type of problem is a very small part of the topic.

Here are two examples of problems you need to be able to solve.

**Example 17:** Solve for $x$: $\log_3 x + \log_3(x + 6) = 3$

**Solution:** By the first law of logs, we get $\log_3 x(x + 6) = 3$. By the definition of logs, $x(x + 6) = 3^3$; $x^2 + 6x - 27 = (x + 9)(x - 3) = 0$. $x$ can't be $-9$ because we can take the log of positive numbers only. The answer is $x = 3$. For fun, let's check: $\log_3 3 + \log_3 9 = 1 + \log_3 3^2 = 1 + 2\log_3 3 = 1 + 2(1) = 3$.

**Example 18:** Solve for $x$: $\log_4 x - \log_4(x - 2) = 2$.

**Solution:** By the second law of logarithms, we get $\log_4 \dfrac{x}{x - 2} = 2$. And by the definition of logs, $4^2 = \dfrac{16}{1} = \dfrac{x}{x - 2}$. By cross-multiplying, we get $16x - 32 = x$; so $x = \dfrac{32}{15}$. As long as this answer does not give the log of a negative number (it doesn't), this is the answer. Otherwise, this problem would have no answer.

**Example 19:** Solve for $x$: $9^{4x + 5} = 6^{2x + 8}$.

**Solution:** Notice that 9 and 6 cannot be taken to the same base. The only way to get the $x$'s off the exponents is to take logs of both sides, to get $(4x + 5)\log 9 = (2x + 8)\log 6$. So $x = \dfrac{8\log 6 - 5\log 9}{4\log 9 - 2\log 6}$.

 **Note** *The algebra can be done in one step!!! Yes, it can. If we multiply out each side, getting all terms without x to the right, and all terms with x to the left, factor out the x, the coefficient of x goes to the bottom of the fraction. You can do it if you try.*

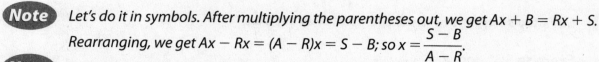

**Note** *Let's do it in symbols. After multiplying the parentheses out, we get $Ax + B = Rx + S$. Rearranging, we get $Ax - Rx = (A - R)x = S - B$; so $x = \dfrac{S - B}{A - R}$.*

**Note** *On a test, I allow the students to leave this answer since pushing buttons on a calculator is silly. However, your teacher or the COMPASS might ask you to do this. If you do, place a set of brackets around the top and a set of brackets around the bottom. On the calculator it is the top bracket divided by the bottom bracket.*

**Note** *Instead of using log, you could have used ln.*

**Exercise 20:** Find $x$.

| Problem | Solution |
|---|---|
| **a.** $\log_{25} x = -\dfrac{3}{2}$ | $\dfrac{1}{125}$ |
| **b.** $\log_x 81 = 4$ | $3$ |
| **c.** $\log_{81} \dfrac{1}{9} = x$ | $-\dfrac{1}{2}$ |
| **d.** $\log_7 7^2 = x$ | $2$ |
| **e.** $\log_9 1 = x$ | $0$ |
| **f.** $e^{\ln 7} = x$ | $7$ |
| **g.** $8^{2\log_8 5} = x$ | $25$ |
| **h.** $\log 5 + \log (2x + 3) = \log 85$ | $7$ |
| **i.** $\log_4\left(\dfrac{x - 3}{2}\right) = 3$ | $131$ |
| **j.** $\log_2 x = 6 - \log_2 4x$ | $4$ |
| **k.** $4^{7x - 5} = 3^{6x - 8}$ | $\dfrac{5\log 4 - 8\log 3}{7\log 4 - 6\log 3}$ |

**(Q)** **Let's try some more multiple-choice exercises.**

**Exercise 5:** $\log_4 x = -2. x =$

A. $-16$

B. $-8$

C. $-\dfrac{1}{16}$

D. $\dfrac{1}{16}$

E. $\dfrac{1}{8}$

**Exercise 6:** $\log_{16} x = \dfrac{-3}{4}. x =$

A. $12$

B. $8$

C. $\dfrac{1}{8}$

D. $-\dfrac{1}{8}$

E. $-\dfrac{1}{12}$

**Exercise 7:** $\log_x 81 = 4. x =$

A. $\sqrt{3}$

B. $3$

C. $9$

D. $\dfrac{81}{4}$

E. There is more than one answer.

**Exercise 8:** $\log_{32} 64 = 5x. x =$

A. $\dfrac{1}{10}$

B. $\dfrac{6}{25}$

C. $\dfrac{2}{5}$

D. $\dfrac{1}{2}$

E. $\dfrac{6}{5}$

**Exercise 9:** $\log_2 x = 5 - \log_2 (x + 14). x =$

A. 2 only

B. 2 and $-16$

C. $-16$ only

D. no real number

E. all real numbers

**Let's look at the answers.**

**Answer 5:**   D: $4^{-2} = \left(\dfrac{1}{4}\right)^2 = \dfrac{1}{16}$.

**Answer 6:**   C: $16^{-3/4} = \dfrac{1}{16^{3/4}} = \dfrac{1}{\left(\sqrt[4]{16}\right)^3} = \dfrac{1}{2^3} = \dfrac{1}{8}$.

**Answer 7:**   B: $x^4 = 81$; $x = 3$; only positive answers are allowed.

**Answer 8:**   B: $32^{5x} = 64$; $(2^5)^{5x} = 2^6$. When the bases are the same, the exponents must also be equal, so $25x = 6$; $x = \dfrac{6}{25}$.

**Answer 9:**   A: This is an incredibly tricky problem until you see the trick. You must put both logs on the same side! $\log_2 x + \log_2 (x + 14) = \log_2 x(x + 14) = 5$; so $x(x + 14) = 2^5 = 32$. Then $x^2 + 14x - 32 = (x + 16)(x - 2) = 0$; so $x = 2$, and $x = -16$, which is not allowed. Every once in a while the COMPASS gives choices for which you can see that only one can possibly be correct, even without doing the problem. From the expression $\log_2 x$ given in the exercise, we know that $x$ must be greater than 0. Immediately, choices B, C, and E must be wrong. Substituting 2 in the problem, we see that it checks. If it didn't check, then the answer had to be "no solution," the only one left.

Logarithms and their inverses, exponentiation (they are inverses, or opposites, similar to adding and subtracting), are the most interesting math-related topics below the level of calculus. Let me tell you the two stories I always tell at this point.

The first involves the Richter scale for earthquakes, which is roughly a logarithmic scale. Did you ever notice that a 3 causes a little shaking, a 5 causes minor damage, and at a 7, all #@$$#& breaks loose. On the Richter scale, to go between two numbers, the power of the earthquake is multiplied by 1,000! A Richter scale 7 is a million times as strong as a Richter scale 3. Hopefully, it won't happen, but an earthquake of magnitude 9 is supposed to hit California before 2030. If it does, the chances are that San Francisco will be gone, Oakland will be gone, the Golden Gate Bridge will be gone. Like in a Superman movie, Nevada might become the West Coast.

Another story: This is one of the few stories for which I tell the punch line first. This is when the little guy wins. In 1905, some guy—I don't remember his name—invented a photographic process of some kind and brought it to a large company. The company said it couldn't use it, and six years later came out with the product. The little guy took the big company to court for stealing his process. Now, normally the little guy loses automatically. Not this time!!! The little guy had taken a picture of San Francisco. As you may know, in 1906 a huge earthquake destroyed San Francisco, and the little guy had a picture of the pre-earthquake city. That is when the little guy wins—when all of San Francisco gets destroyed.

Okay, okay. Enough stories for now.

# EXPONENTIAL GROWTH AND DECAY

This topic is perhaps more fascinating to discuss in a nonmathematical way. It relates to population growth. Assume humans came on the Earth 30,000 years ago (even though the actual figure is at least five times greater!). Count every human who ever lived from 30,000 years ago to the year 1900 (probably 1920 by now). There are more people alive today than that total number. Or go back to 1776. There were 4,000,000 people in the United States. Today there are 8,000,000 people in New York City alone. This shows the tremendous increase in population—exponential growth.

Let's do a little work on exponential growth and decay. We'll do it in two ways, the two ways that the COMPASS might ask. It's still not the way I think it should be done, but it will be enough for now. The first way uses the exponential growth formula as shown in Examples 21 and 22.

The formula for continuous compound interest is given by $A = Pe^{rt}$, where $P$ is the principal, $r$ is the interest rate, $t$ is the time, and $A$ is the amount of money that you earn.

**Example 21:** Find $A$ if we have \$2,800 invested at 6% for 4 years.

**Solution:**  $A = 2,800 \, e^{.06(4)}$. By pushing the proper buttons on the calculator, you would get the answer, which is \$3,559.50.

**Example 22:** Find the interest rate if \$3,000 becomes \$6,000 in 14 years.

**Solution:**  $6,000 = 3,000 \, e^{14r}$; whenever the variable is in the exponent, it is necessary to isolate the term with the variable before taking logs (in this case ln). In this case we divide by 3,000 and get $2 = e^{14r}$. So $\ln 2 = 14r$ and $r = \dfrac{\ln 2}{14}$.

This form of the answer is probably all you need, but if you use your calculator, you will see that this is .05, or 5%.

Now let's look at two examples of the second type of problem.

**Example 23:** Deb started with \$10,000. If her net worth triples every five years, how much will she have in 30 years?

**Solution:**  In 5 years, the \$10,000 will be \$30,000; in 10 years, \$90,000; in 15 years, \$270,000; in 20 years, \$810,000; in 25 years, \$2,430,000; and in 30 years, \$7,290,000!

**Example 24:** A half-life is the time required for half of a radioactive substance to disintegrate. Suppose radioactive goo has a half-life of 15 minutes. If there are 240 pounds of radioactive goo, how much radioactive goo is left in an hour?

**Solution:** In 15 minutes, the 240 pounds will be 120 pounds; in 30 minutes, that 120 pounds will be 60 pounds; in 45 minutes, there will be 30 pounds; and in one hour, only 15 pounds will be left.

**Note** *I teach exponential growth differently from these two methods. However, for this test, this is enough.*

Yesterday I came up with a problem the COMPASS might actually ask. Here it is.

**Example 25:** Solve for $x$: $\ln(4x - 7) + 2 = 19$.

**Solution:** We isolate the log, and then use the definition of logs.
$$\ln(4x - 7) = 17; \text{ so } 4x - 7 = e^{17}; \text{ therefore } x = \frac{e^{17} + 7}{4}.$$

Let's do the chapter test now.

## Chapter 14 Quiz

For questions 1–6, let $f(x) = 5x + 3$, $g(x) = x^2 + 2x + 3$, $k(x) = \dfrac{3x - 5}{2x - 7}$, and $r(x) = \dfrac{1}{x}$.

1. $g(-3) =$

2. $f(g(2)) =$

3. $g(f(x)) =$

4. $k^{-1}(x) =$

5. $k(r(x)) =$

6. $\dfrac{g(x + h) - g(x)}{h} =$

7. Write as simpler logs with no exponents: $\ln\left(\dfrac{pq^6}{5r^7 \cdot \sqrt[3]{s}}\right)$.

8. Write as a single log: $\log 6 + \log 7 + 5\log x - \log e - \log 11$.

9. $3^{4\log_3 5} =$

10. $4\log_5 5^3 =$

11. $\log_7 1^{111} =$

12. If $\log 2 = .30$, $\log 3 = .48$ and $\log 5 = .70$, find $\log_2 3$.

For questions 13–20, solve for x.

13. $\log_x 64 = 6$.

14. $\log_{1/4} x = -\frac{3}{2}$.

15. $\log_{81} 27 = x$.

16. $\log_2 x + \log_2 2x = 5$.

17. $\log_5 x + \log_5(x + 4) = \log_5 21$.

18. $\log_3 x - \log_3(x - 3) = 3$.

19. $9^{3x+2} = 27^{4x-3}$.

20. $4 \log(2x - 6) + 7 = 75$.

For questions 21 and 22, find x; the answers can be in terms of logs, if necessary.

21. $4^{5x-6} = 7^{3x-2}$.

22. $4(10)^{3x+5} = 18$.

For questions 23–25: let $P = P_o(1.02)^t$, where $P_o$ is the population of Earth and t is the number of years. Leave answers in terms of logs or exponentials. Suppose $P_o$ is 6.5 billion people today.

23. How many people will there be in ten years? Do *not* use a calculator.

24. When will there be seven billion people? Answer in terms of ln.

25. In how many years will Earth's population double? Answer in terms of ln.

**Answers to Chapter 14 Quiz**

1. $(-3)^2 + 2(-3) + 3 = 6$.

2. $f(11) = 58$.

3. $(5x+3)^2 + 2(5x+3) + 3 = 25x^2 + 40x + 18$.

4. $k^{-1}(x) = y = \dfrac{3x - 5}{2x - 7}$; so $y(2x - 7) = 2xy - 7y = 3x - 5$; $2xy - 3x = x(2y - 3) = 7y - 5$;

   $x = \dfrac{7y - 5}{2y - 3}$; with the notation change, we get $k^{-1}(x) = \dfrac{7x - 5}{2x - 3}$.

5. $\dfrac{3\left(\frac{1}{x}\right) - 5}{2\left(\frac{1}{x}\right) - 7} = \dfrac{3 - 5x}{2 - 7x}$.

6. $\dfrac{(x + h)^2 + 2(x + h) + 3 - (x^2 + 2x + 3)}{h} = 2x + 2 + h$.

7. $\ln p + 6\ln q - \ln 5 - 7\ln r - \dfrac{1}{3}\ln s.$

8. $\log\left(\dfrac{42x^5}{11e}\right).$

9. $5^4 = 625$

10. $4(3) = 12$

11. $0$

12. $\dfrac{\log 3}{\log 2} = \dfrac{.48}{.30} = 1.6.$

13. $x^6 = 64; x = 2$

14. $\left(\dfrac{1}{4}\right)^{-\frac{3}{2}} = 4^{\frac{3}{2}} = 2^3 = 8.$

15. $81^x = 27; (3^4)^x = 3^3;$ so $4x = 3;$ and $x = \dfrac{3}{4}.$

16. $\log_2 x(2x) = 5; 2x^2 = 32; x = 4$ ($-4$ is no good since you can't have the log of a negative number).

17. $\log_5 x(x+4) = \log_5 21;$ so $x^2 + 4x = 21$ or $x^2 + 4x - 21 = (x+7)(x-3) = 0$ (again, only $x = 3$ because you can't have the log of a negative number).

18. $\log_3\left(\dfrac{x}{x-3}\right) = 3;$ so $\dfrac{x}{x-3} = 3^3 = \dfrac{27}{1};$ cross-multiply to get $27(x-3) = x; x = \dfrac{81}{26}.$

19. $(3^2)^{3x+2} = (3^3)^{4x-3};$ by $1-1$ function, we get $2(3x+2) = 3(4x-3); x = \dfrac{13}{6}.$

20. $\log(2x-6) = 17; 2x-6 = 10^{17}; x = \dfrac{10^{17}+6}{2}.$

21. $(5x-6)\ln 4 = (3x-2)\ln 7; x = \dfrac{6\ln 4 - 2\ln 7}{5\ln 4 - 3\ln 7};$ log is also OK.

22. $10^{3x+5} = 4.5; 3x+5 = \log 4.5; x = \dfrac{\log 4.5 - 5}{3}.$

23. $P = 6.5 \times 10^9 \times (1.02)^{10}.$

24. $7 \times 10^9 = 6.5 \times 10^9 (1.02)^t; \dfrac{14}{13} = 1.02^t; t = \dfrac{\ln\left(\dfrac{14}{13}\right)}{\ln(1.02)}.$

25. $13 \times 10^9 = 6.5 \times 10^9 \times 1.02^t; 2 = 1.02^t;$ so $t = \dfrac{\ln 2}{\ln 1.02}.$ By calculator, Earth's population,

assuming 2% growth annually, doubles every 35 years. Scary! So in 70 years, there will be 26 billion people, and in 105 years, there will be 52 billion of us!!

And now on to trig.

**Answer to "Bob Asks":** The 46th day. After 45 days, the monkey is 45 feet high. On the 46th morning, it climbs 5 feet to 50 feet, getting out of the hole.

"*How much dirt is there in a hole that is 5 feet by 4 feet by 9 inches?*"

**The** secret of making trigonometry (trig) easy—yes, I mean easy—is to draw triangles. One topic, identities, is impossible to relate to triangles and is the only difficult topic in trig. It is treated very, very mildly in any placement test.

Let's start with angles. For trig purposes, we will have two measures of angles. The first one you know. There are 360 degrees in a circle. You may notice that there are many multiples of 60 in our measurements. More than 7,000 years ago, the Babylonians counted in 60s mainly because they thought there were 360 days in a year; 360 is a great number because it has many factors: 1, 2, 3, 4, 5, 6, 8, 9, 10, 12, 15, 18, 20, 24, 30, 36, 40, 45, 60, 72, 90, 120, 180, and 360. Whew!!!

Any intelligent life coming from another world would not have heard of degrees (they would quickly learn, of course!). However, all would have heard of **radians**.

Lay a radius on the circumference of a circle. The angle formed is said to be one radian.

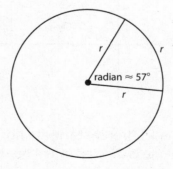

There are a little over 6, or exactly $2\pi$ radians in a circle. So $2\pi = 360°$. This means $1° = \dfrac{\pi}{180}$ radians, and a radian is $\approx 57°$. However, we usually express radians in terms of $\pi$, so we should be concerned with only two conversions:

1.  To change from degrees to radians, multiply by $\dfrac{\pi}{180}$.

2.  To change from radians to degrees, multiply by $\dfrac{180}{\pi}$.

    **Example 1:**   Change $45°$ to radians.

    **Solution:**    $45 \times \dfrac{\pi}{180} = \dfrac{\pi}{4}$ radians.

    **Example 2:**   Change from $\dfrac{\pi}{6}$ radians to degrees.

    **Solution:**    $\dfrac{\pi}{6} \times \dfrac{180}{\pi} = 30°$.

So we see that $\dfrac{\pi}{6}$ radians $= 30°$, and $\dfrac{\pi}{4}$ radians $= 45°$.

It also is a good idea to learn the most important multiples of $30°$ and $45°$:

$\dfrac{\pi}{6} = 30°, \dfrac{\pi}{4} = 45°, \dfrac{\pi}{3} = 60°, \dfrac{\pi}{2} = 90°, \pi = 180°, \dfrac{3\pi}{2} = 270°,$ and $2\pi = 360°$.

If you know more, that is better, but this is enough.

## TRIG RATIOS

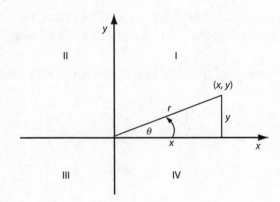

Angles, represented here by the Greek letter theta, $\theta$, are positive if measured counterclock-wise from the positive $x$-axis. The plane is divided into 4 quadrants, as indicated.

Locate the point $(x, y)$ at a distance $r$ from the origin, where $r = \sqrt{x^2 + y^2}$ is always positive.

**Note**  *Always draw the triangle up or down to the x-axis.*

Referring to this triangle, which is a right triangle, we can define the six basic trigonometric ratios:

| Function | Abbreviation | Ratio | Function | Abbreviation | Ratio |
|----------|--------------|-------|----------|--------------|-------|
| sine $\theta$ | sin $\theta$ | $\dfrac{y}{r}$ | cotangent $\theta$ | cot $\theta$ | $\dfrac{x}{y}$ |
| cosine $\theta$ | cos $\theta$ | $\dfrac{x}{r}$ | secant $\theta$ | sec $\theta$ | $\dfrac{r}{x}$ |
| tangent $\theta$ | tan $\theta$ | $\dfrac{y}{x}$ | cosecant $\theta$ | csc $\theta$ | $\dfrac{r}{y}$ |

You must know these definitions!

**Example 3:**   Let $\sin A = \dfrac{7}{10}$ in quadrant II. Find sec $A$.

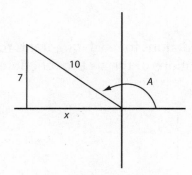

**Solution:**   $\sin A = \dfrac{7}{10} = \dfrac{y}{r}$. Then, to find $x$, we can let $y = 7$ and $r = 10$, and we can use the Pythagorean Theorem: $x = \pm \sqrt{10^2 - 7^2} = \pm \sqrt{51}$. Because $x$ is to the left, it is a negative $x$ value. Then $\sec A = \dfrac{r}{x} = \dfrac{10}{-\sqrt{51}} = \dfrac{-10\sqrt{51}}{51}$.

**Note**  *Once you get x, you have all six of the trig functions!*

**Example 4:**    Let $\tan B = \dfrac{5}{9}$ in quadrant III. Find $\cos B$.

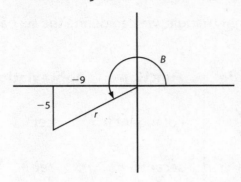

**Solution:**    $\tan B = \dfrac{5}{9} = \dfrac{y}{x}$. In quadrant III, both $x$ and $y$ are negative. We let $y = -5$ and $x = -9$.

Then $r = \sqrt{9^2 + 5^2} = \sqrt{106}$, and $\cos B = \dfrac{x}{r} = \dfrac{-9}{\sqrt{106}} = \dfrac{-9\sqrt{106}}{106}$.

    *Remember, r is always positive.*

It is necessary to know the signs of the trig functions for each quadrant. You should know the following sign diagram. If you know the definitions of the six trig functions, you should be able to figure this out yourself.

|  | II | | I |  |
|---|---|---|---|---|
|  | $x < 0$ | | $x > 0$ |  |
|  | $y > 0$ | | $y > 0$ |  |
|  | $r > 0$ | | $r > 0$ |  |
|  | only sin, csc positive | | all positive |  |
|  | $x < 0$ | | $x > 0$ |  |
|  | $y < 0$ | | $y < 0$ |  |
|  | $r > 0$ | | $r > 0$ |  |
|  | only tan, cot positive | | only cos, sec positive |  |
|  | III | | IV |  |

# FINDING TRIG VALUES FOR MULTIPLES OF 30°, 45°, 60°, AND 90°

For this section, we have to recall the 45°-45°-90° and 30°-60°-90° triangles we discussed in Chapter 10. Let's give specific values for the sides of these triangles.

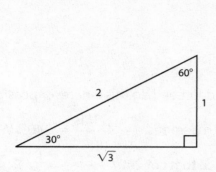

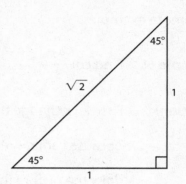

In the 30°-60°-90° triangle, if we let the side opposite the 30° angle = 1, then we know the hypotenuse is twice that, or 2, and we can find the side opposite the 60° angle, $\sqrt{3}$, by using the Pythagorean Theorem.

In the 45°-45°-90° triangle, if we let the sides opposite both 45° angles = 1, the Pythagorean Theorem tells us the hypotenuse, the side opposite the right angle = $\sqrt{2}$.

**Example 5:**    Find cos 150°.

**Solution:**

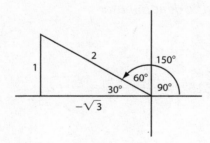

We draw the angle counterclockwise from the positive axis: 90° + 60° = 150°. The angle with the *x*-axis is 30°, giving us a 30°-60°-90° right triangle. The side opposite the 30° angle is 1, and it is +1 because it is "up." The side opposite the 60° angle is $-\sqrt{3}$. It is to the left, so it is negative. The hypotenuse is 2; as we have already seen, it is always positive. So

$$\cos 150° = \frac{x}{r} = \frac{-\sqrt{3}}{2}.$$

**Note**    *We always draw the triangle to the x-axis.*

Let's do two more examples.

**Example 6:** Find $\cot \dfrac{4\pi}{3}$.

**Solution:** First, let's change the angle to degrees. I do this wherever possible because I am more familiar with degrees. $\dfrac{4\pi}{3} \times \dfrac{180}{\pi} = 240°$. When we draw the picture (below), we see that $\cot 240° = \dfrac{x}{y} = \dfrac{-1}{-\sqrt{3}} \times \dfrac{\sqrt{3}}{\sqrt{3}} = \dfrac{\sqrt{3}}{3}$.

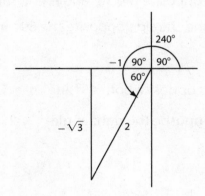

**Note** *The COMPASS always rationalizes the denominator.*

**Example 7:** Find $\csc 315°$.

**Solution:** Draw the figure. $\csc 315° = \dfrac{r}{y} = \dfrac{\sqrt{2}}{-1} = -\sqrt{2}$.

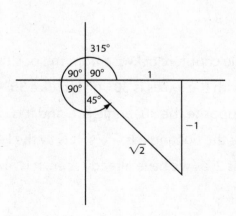

## MULTIPLES OF 90°

To find the values of $x$, $y$, and $r$ for angles on the axes, assume $r$ is along the axis and equals 1. Then determine whether $x$ and $y$ are 0, $+1$, or $-1$ along that axis, and proceed as above.

**Example 8:**    Find all six trig functions for 180°.

**Solution:**

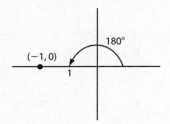

When we draw an angle of 180°, we wind up on the negative $x$-axis. Let $r = 1$, and then the point on the $x$-axis would be $(-1, 0)$ or $x = -1$ and $y = 0$. Once we know $x$, $y$, and $r$, we know all six trig functions. The *COMPASS* would probably ask for only one, but let's do all six in this case.

$$\sin 180° = \frac{y}{r} = \frac{0}{1} = 0 \qquad\qquad \cos 180° = \frac{x}{r} = \frac{-1}{1} = -1$$

$$\tan 180° = \frac{y}{x} = \frac{0}{1} = 0 \qquad\qquad \cot 180° = \frac{x}{y} = \frac{-1}{0}, \text{undefined}$$

$$\sec 180° = \frac{r}{x} = \frac{1}{-1} = -1 \qquad\quad \csc 180° = \frac{r}{y} = \frac{1}{0}, \text{undefined}$$

 *For all multiples of 90°, two of the trig functions will always be 0, two will always be undefined (0 is the denominator), and two will both be either 1 or −1.*

**Example 9:**

| Problem | Solution |
|---|---|
| **a.** Change 50° to radians | $\dfrac{5\pi}{18}$ |
| **b.** Change $\dfrac{11\pi}{12}$ to degrees | 165° |

**c.** Change $7\pi°$ to radians

$$\frac{7\pi^2}{180}$$

**d.** Cos $x = \dfrac{11}{12}$ in quadrant IV; find sin $x$

$$-\frac{\sqrt{23}}{12}$$

**e.** Sin $x = -\dfrac{3}{7}$ in quadrant III; find tan $x$

$$\frac{3\sqrt{10}}{20}$$

**f.** Tan $x = -\dfrac{a}{b}$ in quadrant II; find sec $x$

$$-\frac{\sqrt{a^2 + b^2}}{b}$$

**g.** Cot 120°

$$-\frac{\sqrt{3}}{3}$$

**h.** Csc $\dfrac{7\pi}{4}$

$$-\sqrt{2}$$

**i.** Cos 0°

$$1$$

**j.** Tan $\dfrac{\pi}{2}$

Undefined or none or ∞
(the denominator is 0)

# SKETCHING SINES AND COSINES

You will not be asked to sketch sines and cosines, but some questions can be answered only if you know what the sketches look like. Here, we will do most everything in degrees, but we will refer to radians when appropriate.

All trig functions are **periodic**. That means they repeat. The **period** is the smallest interval in which the trig functions repeat. For sine and cosine (and secant and cosecant), the period is 360° or $2\pi$. The period for tangent and cotangent is 180° or $\pi$.

If we were to graph $y = \sin x$ and $y = \cos x$, the pictures would look like these:

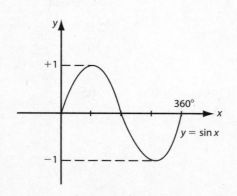

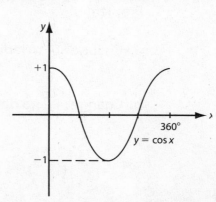

**Example 10:**  Sketch $y = 4 \sin x$.

**Solution:**

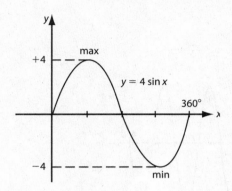

If you draw the figure, it goes four times as high as $y = \sin x$.

**Note**  *The height is called the **amplitude**. The definition of amplitude is* $\dfrac{\text{max} - \text{min}}{2}$.

**Example 11:**  Sketch $y = -6 \cos x$.

**Solution:**

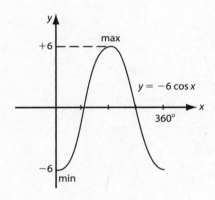

This graph is 6 times as high as $y = \cos x$. The minus sign means the curve is upside down. What is the amplitude? The amplitude is

$$\frac{\text{max} - \text{min}}{2} = \frac{6 - (-6)}{2} = \frac{12}{2} = 6.$$

***Rule 1:*** If $y = A \sin x$ or $y = A \cos x$, the amplitude is $|A|$.

**Example 12:** Sketch $y = 10 \sin 4x$, in degrees.

**Solution:**

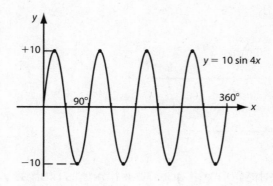

The amplitude is 10, and the graph is not upside down. The graph reaches 360° four times faster, so the period is $\dfrac{360°}{4} = 90°$.

**Example 13:** Sketch $y = -7 \sin \dfrac{x}{5}$.

**Solution:**

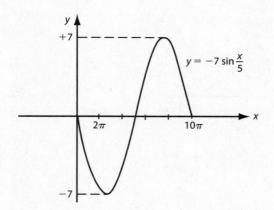

The amplitude is 7, and the graph is upside down. The period is $\dfrac{2\pi}{\frac{1}{5}} = 10\pi$, meaning the graph is "stretched out" five times more than $y = \sin x$.

So we can add to Rule 1:

**Rules 1 and 2:** If $y = A \sin Bx$ or $y = A \cos Bx$, the amplitude is $|A|$, and the period is $\dfrac{360°}{B}$, or $\dfrac{2\pi}{B}$ radians.

**Example 14:**  Sketch $y = -2 \sin (3x - 120°)$.

**Solution:**

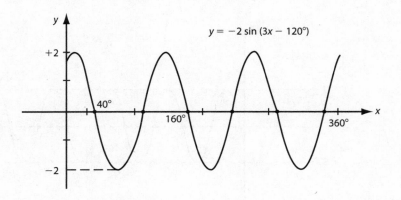

The amplitude is 2, the graph is upside down, and the period is $\dfrac{360°}{3} = 120°$. We know $\sin 0° = 0$, so we set $3x - 120° = 0°$, and we get $x = 40°$. This means a **left-right shift** of $y = 0$ is 40° to the right.

**Example 15:**  Sketch $y = 9 \cos (6x + \dfrac{\pi}{5})$.

**Solution:**

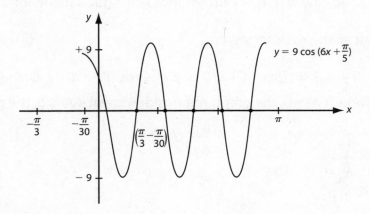

The amplitude is 9, the graph is not upside down, and the period is $\dfrac{2\pi}{6} = \dfrac{\pi}{3}$. We know $\cos 0°$ is at its maximum, so we set $6x + \dfrac{\pi}{5} = 0$, $x = -\dfrac{\pi}{30}$, so there is a $\dfrac{\pi}{30}$ shift of the maximum to the left.

So we can add another rule:

**Rules 1, 2, and 3:**  If $y = A \sin (Bx + C)$ or $y = A \cos (Bx + C)$, the amplitude is $|A|$, the period is $\dfrac{360°}{B}$, or $\dfrac{2\pi}{B}$ , and the left-right shift is $-\dfrac{C}{B}$.

**Example 16:**  Sketch $y = 2 \sin (5x - 30°) + 7$.

**Solution:**

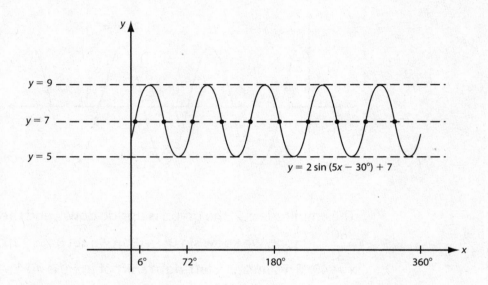

The amplitude is 2, the graph is not upside down, the period is $\dfrac{360°}{5} = 72°$, the left-right shift is $-\dfrac{(-30°)}{5} = 6°$ to the right, and the up-down shift is 7 up (sounds like a good name for a soft drink!).

So finally, we can put all the rules together:

**Rules 1, 2, 3, and 4:**  If $y = A \sin (Bx + C) + D$ or $y = A \cos (Bx + C) + D$, then

- $|A|$ is the amplitude. If $A > 0$, the graph not upside-down; if $A < 0$, the graph is upside down.

- Period is $\dfrac{360°}{B}$ or $\dfrac{2\pi}{B}$.

- Left-right shift is $-\dfrac{C}{B}$.

- Up-down shift is $D$: if $D > 0$, it is up; if $D < 0$, it is down.

That's it for sketching sines and cosines.

# INVERSE TRIG FUNCTIONS

There is no perfect place to put this subject, so let's put it here.

First we need a brief review, as I always do. To have an inverse trig function, we need an inverse function. Before we have an inverse function we need a function. In order to have a function, a graph must pass the vertical line test. To have an inverse function, a graph must also past the horizontal line test. We just drew sine curves, which always pass the vertical line test. However, if you tried a horizontal line test, each line would hit the graph an infinite number of times. So we must restrict the original angles that the trig functions can take on. Also remember that in inverses, the range and domain of the original functions switch.

The *notation* for the inverse of sine is $\sin^{-1}A$ or arc sin $A$, and it means "the angle whose sine is $A$."

Typically, you should know the range for the inverses of sine, cosine, and tangent. Here is a list of their domains and ranges.

| Function | Domain | Range |
|---|---|---|
| $\theta = \sin^{-1}A$ | $-1 \leq A \leq 1$ | $-\dfrac{\pi}{2} \leq \theta \leq \dfrac{\pi}{2}$, or $-90° \leq \theta \leq 90°$ |
| $\theta = \cos^{-1}A$ | $-1 \leq A \leq 1$ | $0 \leq \theta \leq \pi$, or $0° \leq \theta \leq 180°$ |
| $\theta = \tan^{-1}A$ | $-\infty < A < \infty$ | $-\dfrac{\pi}{2} < \theta < \dfrac{\pi}{2}$, or $-90° < \theta < 90°$ |

Inverse problems have only one answer. The positive inverse functions are all in quadrant I. Inverse cosine is in quadrant II, and inverse sine and inverse tangent are in quadrant IV.

The secret (*shh*) of doing inverse trig problems is to read it properly. The answer is an angle.

So reading $\sin^{-1}\left(\dfrac{1}{2}\right)$ as "The angle whose sine is $\dfrac{1}{2}$," we can draw the triangle in quadrant I. We see that the angle is 30°, or $\dfrac{\pi}{6}$.

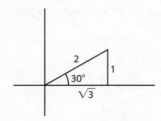

 *The answer should always be in radians, at least when I taught it. However, we understand degrees better, so I put both!*

**Example 17:** $\sin^{-1}\left(-\dfrac{1}{2}\right) =$

**Solution:** This is read as "Find the angle whose sine is minus one-half." The angle is in quadrant IV. Drawing the triangle, we see the answer is $-30°$, or $\left(-\dfrac{\pi}{6}\right)$.

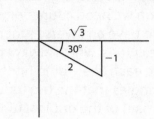

**Note** *Important! The answer is not 330°, or $\dfrac{11\pi}{6}$, since that angle is not in our acceptable range of $-90° \le \theta \le 90°$.*

**Example 18:** Find $\tan\left(\cos^{-1}\left(\dfrac{5}{8}\right)\right).$

**Solution:** This is read as, "Find the tangent of the angle whose cosine is $\dfrac{5}{8}$." You can almost see the picture in your mind; $x = 5$ and $r = 8$; so $y = \sqrt{8^2 - 5^2} = \sqrt{39}$. Tangent $\theta = \dfrac{y}{x} = \dfrac{\sqrt{39}}{5}$.

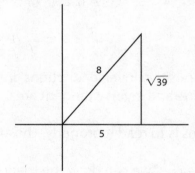

**Example 19:** Find $\sin\left(\tan^{-1}\left(\dfrac{q}{s}\right)\right).$

**Solution:** We draw the triangle with $y = q$ and $x = s$. The hypotenuse is $r = \sqrt{q^2 + s^2}$. Since the sine is $\dfrac{y}{r}$, the answer is $\dfrac{q}{\sqrt{q^2 + s^2}}$.

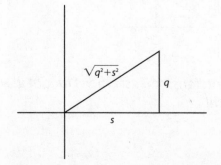

# TRIGONOMETRIC IDENTITIES

An **identity** is an equation that is always true as long as it is defined. For example, $2x + 3x = 5x$ is an identity because no matter what number is used for $x$, $2x + 3x = 5x$.

You should know some trigonometric identities. Let's rewrite the basic six ratios again:

$$\sin \theta = \frac{y}{r} \qquad \cos \theta = \frac{x}{r} \qquad \tan \theta = \frac{y}{x}$$

$$\cot \theta = \frac{x}{y} \qquad \sec \theta = \frac{r}{x} \qquad \csc \theta = \frac{r}{y}$$

Now let's list the eight identities you absolutely, positively should know perfectly!

| Identity | Reason |
|---|---|
| 1. $\sin A \times \csc A = 1$ | $\left(\dfrac{y}{r}\right)\left(\dfrac{r}{y}\right) = 1$ |
| 2. $\cos A \times \sec A = 1$ | $\left(\dfrac{x}{r}\right)\left(\dfrac{r}{x}\right) = 1$ |
| 3. $\tan A \times \cot A = 1$ | $\left(\dfrac{y}{x}\right)\left(\dfrac{x}{y}\right) = 1$ |
| 4. $\tan A = \dfrac{\sin A}{\cos A}$ | $\dfrac{y}{r} \div \dfrac{x}{r} = \dfrac{y}{r} \times \dfrac{r}{x} = \dfrac{y}{x}$ |
| 5. $\cot A = \dfrac{\cos A}{\sin A}$ | $\dfrac{x}{r} \div \dfrac{y}{r} = \dfrac{x}{r} \times \dfrac{r}{y} = \dfrac{x}{y}$ |
| 6. $\sin^2 A + \cos^2 A = 1$ | Divide each term of $x^2 + y^2 = r^2$ by $r^2$. |
| 7. $1 + \tan^2 A = \sec^2 A$ | Divide each term of $x^2 + y^2 = r^2$ by $x^2$. |
| 8. $1 + \cot^2 A = \csc^2 A$ | Divide each term of $x^2 + y^2 = r^2$ by $y^2$. |

**Let's do a few multiple-choice exercises.**

**Exercise 1:**    $\sin G \times \cot G =$

A. $\sin G$              D. $\cot G$

B. $\cos G$              E. $\sec G$

C. $\tan G$

**Exercise 2:**    Given $\sin (A + B) = \sin A \cos B + \cos A \sin B$, and given $\dfrac{5\pi}{12} = \dfrac{\pi}{4} + \dfrac{\pi}{6}$,

then $\sin \left( \dfrac{5\pi}{12} \right) =$

A. $\dfrac{1}{2}$                      D. $\dfrac{\sqrt{6} + \sqrt{2}}{4}$

B. $\dfrac{\sqrt{3}}{2}$                    E. $1$

C. $\dfrac{\sqrt{6} - \sqrt{2}}{4}$

**A**  **Let's look at the answers.**

**Answer 1:**    B: $\cot G = \dfrac{\cos G}{\sin G}$, so $\sin G \times \dfrac{\cos G}{\sin G} = \cos G$.

**Answer 2:**    D: First change each of the radians to degrees. We get $75° = 45° + 30°$.
So $\sin 75° = \sin 45° \cos 30° + \cos 45° \sin 30°$. Then $\sin(45° + 30°) =$

$$\frac{1}{\sqrt{2}} \times \frac{\sqrt{3}}{2} + \frac{1}{\sqrt{2}} \times \frac{1}{2} = \frac{\sqrt{3} + 1}{2\sqrt{2}} \times \frac{\sqrt{2}}{\sqrt{2}} = \frac{\sqrt{6} + \sqrt{2}}{4}.$$

You can get the same answer without changing radians to degrees, if you are comfortable working with radians.

## TRIGONOMETRIC EQUATIONS

Trig equations are like regular equations except we must find the angles. The following three examples will give answers in both degrees and radians, with all answers between $0°$ and $360°$ $(0$ and $2\pi)$ including $0°$.

**Example 20:** Solve for $x$: $\cos x + 2 \sin x \cos x = 0$.

**Solution:** Take out the common factor: $\cos x (2 \sin x + 1) = 0$. This means either $\cos x = 0$ or $\sin x = -\dfrac{1}{2}$. We know sine is negative in quadrants III and IV. We then draw the appropriate triangles and we see that $x = 210°$ and $330°$, or in radians, $\dfrac{7\pi}{6}$ and $\dfrac{11\pi}{6}$.

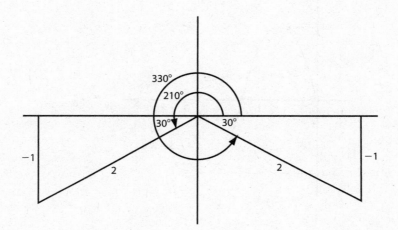

A little trick: Draw the $y = \cos x$ curve. From this curve, we see that $\cos x = 0$ at $90°$ or $270°$, or in radians, $\dfrac{\pi}{2}$ or $\dfrac{3\pi}{2}$. So the answers to Example 20 are $x = 90°, 210°, 270°,$ and $330°$, or $x = \dfrac{\pi}{2}$, $\dfrac{7\pi}{6}, \dfrac{3\pi}{2},$ and $\dfrac{11\pi}{6}$ radians.

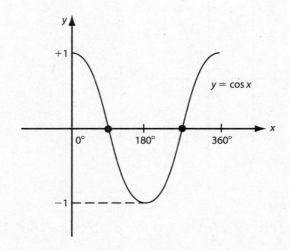

**Note** *Also note that $\cos 0° = \cos 360° = 1$, and $\cos 180° = -1$.*

**Example 21:** Solve for $x$: $4 \sin^3 x - 3 \sin x = 0$.

**Solution:** We get $\sin x (4 \sin^2 x - 3) = 0$. Thus, $\sin x = 0$ or $4 \sin^2 x = 3$; $\sin^2 x = \dfrac{3}{4}$; $\sin x = \pm \dfrac{\sqrt{3}}{2}$. $\sin x = 0$ for $x = 0°$ and $180°$ (see the sine curve below), or $0$ and $\pi$ radians. We don't include $360°$ or $2\pi$ because it is the same as $0$.

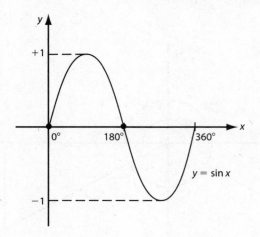

Next, we see $\sin x = + \dfrac{\sqrt{3}}{2}$ in quadrants I and II, and $x = - \dfrac{\sqrt{3}}{2}$ in quadrants III and IV. From the graph below, we can see that $x = 60°$, $120°$, $240°$, and $300°$, or in radians, $x = \dfrac{\pi}{3}, \dfrac{2\pi}{3}, \dfrac{4\pi}{3}$, and $\dfrac{5\pi}{3}$.

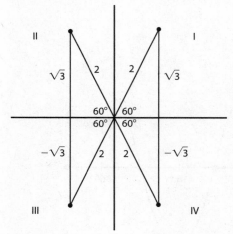

There are six solutions to this problem: $0°$, $60°$, $120°$, $180°$, $240°$, and $300°$, or $0, \dfrac{\pi}{3}, \dfrac{2\pi}{3}, \pi, \dfrac{4\pi}{3}$, and $\dfrac{5\pi}{3}$ radians.

 $\sin 90°$ (or $\sin \dfrac{\pi}{2}$) $= 1$; $\sin 270°$ (or $\sin \dfrac{3\pi}{2}$) $= -1$.

**Example 22:**    Solve for $x$: $\tan^2 x + \sec x - 1 = 0$.

**Solution:**    We let $\tan^2 x = \sec^2 x - 1$. The equation becomes $\sec^2 x - 1 + \sec x - 1$

$= \sec^2 x + \sec x - 2 = (\sec x + 2)(\sec x - 1) = 0$. So the roots of the

equation are $\sec x = -2$ or $\sec x = 1$. But $\cos x = \dfrac{1}{\sec x}$, so we can

rewrite the roots as $\dfrac{1}{\cos x} = -2$, or $\cos x = -\dfrac{1}{2}$ and $\cos x = 1$.

From the graph of $y = \cos x$, we see that $\cos x = 1$ means $x = 0°$, or

0 radians. For $\cos x = -\dfrac{1}{2}$, $\cos x$ is negative in quadrants II and III, as seen

in the graph below. Then $x = 120°$ and $240°$, or $\dfrac{2\pi}{3}$ and $\dfrac{4\pi}{3}$ radians.

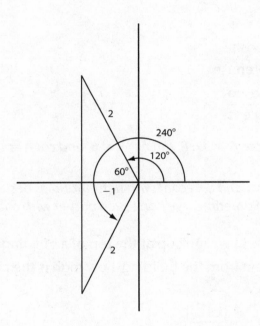

The answers to Example 22 are $0°$, $120°$, and $240°$, or $0$, $\dfrac{2\pi}{3}$, and

$\dfrac{4\pi}{3}$ radians.

## RIGHT-ANGLE TRIGONOMETRY

I believe that if you do non-right-angle trig first and then do right-angle trig last, right-angle trig seems very easy. If you do right-angle trig first, then non-right-angle trig is not so easy. That is why I'm following this order.

Take the definitions we have been using. Put the triangle in the first quadrant. Then remove the triangle from the $x$ and $y$ axes. You get the following right triangle:

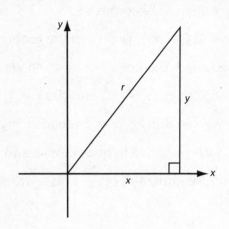 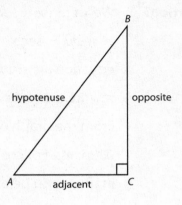

Notice:   $\sin A = \dfrac{y}{r} = \dfrac{BC}{AB} = \dfrac{\text{opposite}}{\text{hypotenuse}}$          $\cos A = \dfrac{x}{r} = \dfrac{AC}{AB} = \dfrac{\text{adjacent}}{\text{hypotenuse}}$

$\tan A = \dfrac{y}{x} = \dfrac{BC}{AC} = \dfrac{\text{opposite}}{\text{adjacent}}$          $\cot A = \dfrac{x}{y} = \dfrac{AC}{BC} = \dfrac{\text{adjacent}}{\text{opposite}}$

**Note**   *Notice that $\sin A = \cos B$, $\cos A = \sin B$, $\tan A = \cot B$, and $\cot A = \tan B$.*

**Note**   *We don't bother with secant and cosecant with right triangles because no one does in the Age of the Calculator. Sometimes we don't even bother with cotangent.*

**Example 23:**   From the ground we look up at the top of a building at an angle of 53°, if we are 34 feet from the building, how high is the building?

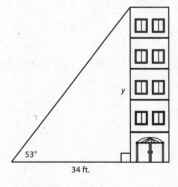

**Solution:**   From the picture, we get that $\tan 53° = \dfrac{y}{34}$; $y = 34 \tan 53°$. Tan 53° ≈ 1.33, so $y \approx 45$ feet.

**Note**   *This angle is called the **angle of elevation**.*

**Example 24:** From the top of a lighthouse 360 feet high, the **angle of depression** to a boat is 37°. How far is the boat from the bottom of the lighthouse, and how far from you is the boat?

**Note** *The **angle of depression** is the angle between a line parallel to the horizon and the object.*

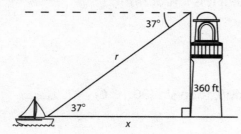

**Solution:** From the picture, by alternate interior angles (which are equal), the angle of depression equals the angle of elevation. So we have

$$\tan 37° = \frac{360}{x}; \text{ so } x = \frac{360}{\tan 37°} \approx \frac{360}{0.75} = 480 \text{ feet from boat to light}$$

house base.

Now, $\sin 37° = \frac{360}{r}$, so $r = \frac{360}{\sin 37°} \approx \frac{360}{0.60} = 600$ feet from boat to top of light house.

**Note** *If you have two sides missing in the same problem, always use a trig function that involves the side you are given, for three reasons: it is exactly correct (it is given); you may have made a mistake in the first part; and even if you are correct, the new side part you obtained is approximately correct.*

**Example 25:** In a 5–12–13 right triangle, find all the angles.

**Solution:** We know one angle is 90°! The tangent of the smallest angle is given by $\tan x = \frac{5}{12}$. So $x = \tan^{-1}\left(\frac{5}{12}\right)$. By calculator, this is approximately 23°. The third angle is then equal to 90° − 23°, or approximately 67°.

## Chapter 15 Quiz

1. Csc 315° =

2. Tan $\dfrac{7\pi}{6}$ =

3. Cot 180° =

For questions 4–7, let $y = -10 \sin\left(6x - \frac{\pi}{3}\right) - 30$.

4. What is the amplitude?

5. What is the period?

6. What is the left-right shift? Indicate if it is left or right.

7. What is the up-down shift?

8. What is the value of $\dfrac{\cos^2 x}{1 - \sin^2 x}$?

9. If $\sin(A - B) = \sin A \cos B - \cos A \sin B$, then $\sin(90° - C) = $ _____

10. $\cos 150° = $ _____

11. A person looks up at the top of a 100-foot tree at an angle of 53°. How far from the tree is the person?

12. $\cos^{-1}\left(\dfrac{-1}{\sqrt{2}}\right) = $

13. $\tan\left(\sin^{-1}\left(\dfrac{2}{3}\right)\right) = $

14. If $m$ and $n$ are in the quadrant I, $\csc\left(\cot^{-1}\left(\dfrac{m}{n}\right)\right) = $

15. Solve for $x$, $0 \le x < 360°$ (also write angles in radians): $4\cos^3 x - \cos x = 0$.

16. Solve for $x$, $0 \le x < 360°$ (also write angles in radians): $\sqrt{3}\tan^2 x + \tan x = 0$.

17. $\sin(A + B) = \sin A \cos B + \cos A \sin B$. Find $\sin 105°$.

18. If $\cos\dfrac{A}{2} = \pm\sqrt{\dfrac{1 + \cos A}{2}}$, find $\cos 165°$.

19. If $\cos 2A = \cos^2 A - \sin^2 A$, find $\cos 2A$ if $\cos A = \dfrac{4}{b}$, where $A$ is in quadrant I.

20. Show that $\sec^2 A + \csc^2 A = \sec^2 A \times \csc^2 A$.

## Answers to Chapter 15 Quiz

1. $-\sqrt{2}$.

2. $\dfrac{\sqrt{3}}{3}$.

3. Undefined.

4. 10.

5. $\dfrac{2\pi}{6} = \dfrac{\pi}{3}$.

6. $6x - \dfrac{\pi}{6} = 0$; $6x = \dfrac{\pi}{3}$; $x = \dfrac{\pi}{3} \times \dfrac{1}{6} = \dfrac{\pi}{18}$, right shift.

7. Down 30.

8. $\dfrac{\cos^2 x}{1 - \sin^2 x} = \dfrac{\cos^2 x}{\cos^2 x} = 1$.

9. $(\sin 90°)(\cos C) - (\cos 90°)(\sin C)$

   $= (1)(\cos C) - (0)(\sin C) = \cos C$.

10. $\cos (150°) = -\cos (30°) = -\dfrac{\sqrt{3}}{2}$.

11. $\text{Tan } 53° = \dfrac{100}{x}$; $x = \dfrac{100}{\tan 53°} \approx 75$ feet.

12. If the angle is in the second quadrant; 135°, or $\dfrac{3\pi}{4}$. If the angle is in the third quadrant, 225°, or $\dfrac{5\pi}{4}$.

13. Let $y = 2$, $r = 3$; then $x = \pm \sqrt{3^2 - 2^2} = \pm\sqrt{5}$; $\tan = \dfrac{y}{x} = \dfrac{2}{\pm\sqrt{5}} = \dfrac{\pm 2\sqrt{5}}{5}$.

14. We let $x = m$ and $y = n$; so $r = \sqrt{m^2 + n^2}$; cosecant $= \dfrac{r}{y} = \dfrac{\sqrt{m^2 + n^2}}{n}$.

15. By factoring, we get $\cos x(2\cos x - 1)(2 \cos x + 1) = 0$. $\cos x = 0$ at 90° and 270° $\left(\text{or } \dfrac{\pi}{2} \text{ and } \dfrac{3\pi}{2} \text{ radians}\right)$ from the graph of $y = \cos x$. $\cos x = \dfrac{1}{2}$ in quadrants I and IV, so $x = 60°$ and 300° $\left(\text{or } \dfrac{\pi}{3} \text{ and } \dfrac{5\pi}{3} \text{ radians}\right)$. $\cos x = -\dfrac{1}{2}$ in quadrants II and III; so $x = 120°$ and 240° $\left(\text{or } \dfrac{2\pi}{3} \text{ and } \dfrac{4\pi}{3} \text{ radians}\right)$.

16. By factoring, we get $\tan x\left(\sqrt{3} \tan x + 1\right)$. $\tan x = 0$ if $x = 0°$ or 180° (or 0 or $\pi$ radians) (note that 360° is not included). $\tan x = -\dfrac{1}{\sqrt{3}}$ in quadrants II and IV; so $x = 150°$ and 330°, $\left(\text{or } \dfrac{5\pi}{6} \text{ and } \dfrac{11\pi}{6} \text{ radians}\right)$.

17. $\text{Sin } 105° = \sin (60° + 45°) = \sin (60°) \cos (45°) + \cos 60° \sin (45°) = \dfrac{\sqrt{6} + \sqrt{2}}{4}$.

18. $\text{Cos } 165° = \pm \sqrt{\dfrac{1 + \cos 330°}{2}}$; since 165° is in the second quadrant, the cosine is negative, so we use the minus sign. The final (ugly) answer is $-\sqrt{\dfrac{1 + \dfrac{\sqrt{3}}{2}}{2}}$, or $\dfrac{-\sqrt{2 + \sqrt{3}}}{2}$.

19. Since $\sin^2 A + \cos^2 A = 1$ and $A$ is in the first quadrant, $\sin^2 A = 1 - \cos^2 A =$
    $1 - \left(\dfrac{4}{b}\right)^2 = \dfrac{b^2 - 16}{b^2}$; so $\cos 2A = \cos^2 A - \sin^2 A = \dfrac{16}{b^2} - \dfrac{b^2 - 16}{b^2} = \dfrac{32 - b^2}{b^2}$.

20. Working only with the left side, we get
    $$\dfrac{1}{\cos^2 A} + \dfrac{1}{\sin^2 A} = \dfrac{\cos^2 A + \sin^2 A}{\cos^2 A \times \sin^2 A} = \dfrac{1}{\cos^2 A \times \sin^2 x} = \dfrac{1}{\cos^2 x} \times \dfrac{1}{\sin^2 x} =$$
    $\sec^2 A \times \csc^2 A$. It's a really weird identity; you have two trig functions; whether you add them or multiply them, you get the same answers, always, for any angle!

That's all for the trig. Let's get to the chapter on the topics that didn't seem to fit well into any other chapter.

**Answer to "Bob Asks":** There is no dirt in a hole.

# CHAPTER 16: *Topics That Don't Fit in Anywhere Else*

"*An ice cream cake has a radius of 7 inches and costs $10.00. If the radius is doubled, how much should the cake cost if the owner charges the same rate? (The height is the same.)*"

**The** title of this chapter is not quite accurate. It is a chapter that contains material for which, if I wanted to place my students properly, not a single question would be asked. It's not that the math in this chapter is unimportant or unnecessary. The fact is that the placement exam is not 79 hours long, or the school is not asking you to answer a 457-question test. Multiple choice doesn't show how well you know a question, but when you are testing many students quickly, it is the fastest way. If I have only a few questions to find out how well you know your math, there are specific skills I want to see. None are from this chapter. However, you and I are both slaves to the system. So here are the topics that don't fit anywhere else.

Let's start at the beginning, counting.

## COUNTING

The **basic law of counting** says: "If you can do something in $p$ ways, and a second thing in $q$ ways, and a third thing in $r$ ways, and so on, the total number of ways you can do the first thing, then the second thing, then the third thing , etc., is $p \times q \times r \times \dots$.

> **Example 1:** If we have a lunch choice of 5 sandwiches, 4 desserts, and 3 drinks, and we can have one of each, how many different meals could we choose?
>
> **Solution:** We can choose from $(5)(4)(3) = 60$ different meals

### Arrangements

Let $n(A)$ be the number of elements in set $A$. In how many ways can these elements be arranged? The answer is that the first has $n$ choices, the second has $(n - 1)$ choices (because one is already used), the third has $(n - 2)$ choices, all the way down to the last element, which has only one choice. In general, if there are $n$ choices, the number of ways to choose is $n!$ (read as "**n factorial**") $= n(n - 1)(n - 2) \times \dots (3)(2)(1)$.

**Example 2:**   In how many ways can five people line up?

**Solution:**   This is just $5 \times 4 \times 3 \times 2 \times 1 = 120$.

**Note**   *Write $5 \times 4 \times 3 \times 2 \times 1$ as 5!, read "5 factorial." In addition, 0! = 1 since this definition makes all the formulas true! (Not factorial.)*

**Example 3:**   How many ways can 5 people sit in a circle?

**Solution:**   It would appear to be the same question as Example 2, but it's not. If we draw the picture, each of the five positions would be the same. The answer is $(5)(4)(3)(2)(1) \div 5 = (4)(3)(2)(1) = 24$. So $n$ people can sit in a circle in $(n - 1)!$ ways.

## Permutations

**Permutations** are essentially the law of counting without repeating, but order is important.

**Example 4:**   How many ways can 7 people occupy 3 seats on a bench?

**Solution:**   Any one of 7 people can be in the first seat, then any one of 6 people can be in the second seat, and any one of 5 people can be in the third seat. The total number would be $(7)(6)(5) = 210$ ways. There are many notations for permutations. One notation for this example would be $P(7, 3)$. Another symbol is $P_{7,3}$.

In general $P(n, r) = \dfrac{n!}{(n - r)!}$.

So $P(7, 3) = \dfrac{7!}{(7 - 3)!} = \dfrac{7 \times 6 \times 5 \times 4 \times 3 \times 2 \times 1}{4 \times 3 \times 2 \times 1} = 7 \times 6 \times 5$.

## Combinations

**Combinations** are essentially the law of counting, with no repetition, and order doesn't matter.

**Example 5:**   How many sets of three different letters can be made from eight different letters?

**Solution:**   Because order doesn't matter, unlike the case for permutations, $AB$ is the same as $BA$. So we can take the number of permutations, but we have to divide by the number of duplicates. It turns out that the duplicates for 3 letters is $3 \times 2 \times 1 = 6$. So we would have $\dfrac{8 \times 7 \times 6}{3 \times 2 \times 1} = 56$.

Again, there are many notations for combinations; the most common are $C(n, r)$, $C_{n,r}$, and $\binom{n}{r}$.

In general, $C(n, r) = \dfrac{n!}{r!(n - r)!}$.

So $C(8, 3) = \dfrac{8!}{3!5!} = \dfrac{8 \times 7 \times 6 \times 5 \times 4 \times 3 \times 2 \times 1}{3 \times 2 \times 1 \times 5 \times 4 \times 3 \times 2 \times 1} = \dfrac{8 \times 7 \times 6}{3 \times 2 \times 1} = 56.$

Usually, when the words "set" or "committee" are used, it implies combinations. If an element is in a set, order doesn't matter. If you are on a committee, the order doesn't matter. You're stuck on the committee.

## Avoiding Duplicates

When we count how many ways to do $A$ or $B$, we should be careful not to count any item twice. We must subtract any items that include both $A$ and $B$:

$$N(A \text{ or } B) = N(A) + N(B) - N(A \text{ and } B)$$

**Example 6:** Thirty students take French or German. If 20 took French and 18 took German, and if each student took at least one language, how many took both French and German?

**Solution:** $N(A \text{ or } B) = N(A) + N(B) - N(A \text{ and } B)$ or $30 = 20 + 18 - x$, so $x = 8$ took both languages.

**Example 7:** Forty students take Chinese or Japanese. If 9 take both and 20 take Japanese, how many students take Chinese?

**Solution:** $N(C \text{ or } J) = N(C) + N(J) - N(\text{both})$, or $40 = x + 20 - 9$, so $x = 29$ take Chinese.

**Q** **Let's do some multiple-choice exercises.**

For Exercises 1–5, use the set $\{e, f, g, h, i\}$. A word is considered to be any group of letters together; for example, hhg is a three-letter word.

**Exercise 1:** From this set, the number of three-letter words is:

A. 6      D. 60

B. 27      E. 125

C. 30

**Exercise 2:**    How many three-letter permutations are there in this set?

　　　A. 6                          D. 60

　　　B. 27                         E. 125

　　　C. 30

**Exercise 3:**    How many three-letter words starting with a vowel and ending in a consonant can be made from this set?

　　　A. 6                          D. 60

　　　B. 27                         E. 125

　　　C. 30

**Exercise 4:**    How many three-letter words with the second and third letters the same can be made from this set?

　　　A. 5                          D. 60

　　　B. 20                         E. 125

　　　C. 25

**Exercise 5:**    How many three-letter permutations with the first and last letters *not* vowels can be made from this set?

　　　A. 18                         D. 45

　　　B. 27                         E. 125

　　　C. 30

**Exercise 6:**    Fifty students take Spanish or Portuguese. If 20 take both and 40 take Spanish, the number of students taking Portuguese *only* is

　　　A. 0                          D. 20

　　　B. 5                          E. 30

　　　C. 10

Ⓐ  **Let's look at the answers.**

**Answer 1:**    E: (5)(5)(5) = 125.

**Answer 2:**    D: (5)(4)(3) = 60.

**Answer 3:**   C: The first letter has 2 choices, the second can be any of 5, and the third has 3 choices, so (2)(5)(3) = 30.

**Answer 4:**   C: There are 5 choices for the first two letters, but there is only 1 choice for the third letter because it must be the same as the second, so (5)(5)(1) = 25.

**Answer 5:**   A: There are 3 choices for the first letter, but only 2 choices for the last letter because it can't be a vowel and must be different than the first letter. There are three choices for the middle letter because two letters have already been used, so the answer is (3)(3)(2) = 18. These questions must be read very carefully!

**Answer 6:**   C: This is not quite the same as the previous exercise. $N(S$ or $P) = N(S) + N(P) - N(both); 50 = 40 + x - 20; x = 30$. But that is not the answer. If 30 take Portuguese and 20 take both, then 10 take Portuguese only.

# PROBABILITY

The probability of an event is the number of "good" outcomes divided by the total number of outcomes possible, or $Pr(success) = \dfrac{\text{good outcomes}}{\text{total outcomes}}$.

**Example 8:**   Consider the following sets: {26-letter English alphabet}; vowels = {$a, e, i, o, u$}; consonants = {the rest of the letters}. What are the probabilities of choosing a vowel? a consonant? any letter? $\pi$?

**Solutions:**   $Pr(\text{vowel}) = \dfrac{5}{26}; Pr(\text{consonant}) = \dfrac{21}{26}; Pr(\text{letter}) = \dfrac{26}{26} = 1; Pr(\pi) = \dfrac{0}{26} = 0.$

Probability follows the same rule about avoiding duplicates as discussed in the previous section.

$$Pr(A \text{ or } B) = Pr(A) + Pr(B) - Pr(A \text{ and } B)$$

**Example 9:**   What is the probability that a spade or an ace is pulled from a 52-card deck?

**Solution:**   $Pr(\text{Spade or ace}) = Pr(\text{Spade}) + Pr(\text{Ace}) - Pr(\text{Spade ace}) =$

$\dfrac{13}{52} + \dfrac{4}{52} - \dfrac{1}{52} = \dfrac{16}{52} = \dfrac{4}{13}.$

As weird as it sounds, whenever I taught this in a class, I never failed to have at least two students who didn't know what a deck of cards was, and I taught in New York City!

Use this figure for Examples 10 and 11. In the jar are five red balls and three yellow balls.

**Example 10:**   What is the probability that two yellow balls are picked, with replacement?

**Solution:**     $Pr$(2 yellow balls, with replacement) $= \left(\dfrac{3}{8}\right)\left(\dfrac{3}{8}\right) = \dfrac{9}{64}$

**Example 11:**   What is the probability of picking two yellow balls, without replacement?

**Solution:**     $Pr$(2 yellow balls, no replacement) $= \left(\dfrac{3}{8}\right)\left(\dfrac{2}{7}\right) = \dfrac{3}{28}$

## SETS

We have mentioned the topic of sets informally. Now let's be a little more formal.

Sets are denoted by braces $\{a, b, c\}$. This set has three elements: $a$, $b$, and $c$.

**Example 12:**   Let $A = \{a, b\}$, $B = \{b, a\}$, $C = \{a, b, b, b, a, b, b, a, a\}$. How is set $A$ related to set $B$? To set $C$?

**Solutions:**    $A = B$ because the order does not matter in sets. $A = C$ because repeated elements are counted only once; $C$ has only two elements in it: $a$ and $b$.

We write $A \cup B$, read "$A$ **union** $B$," as the set of elements in $A$ or in $B$ or in both.

We write $A \cap B$, read "$A$ **intersect(ion)** $B$," as the set of elements common to $A$ and $B$.

The **null set**, written $\{\}$ or $\phi$ (the Greek letter phi), is the set with nothing in it.

**Example 13:**   Let $D = \{a, b, c, d, e, f\}$, $E = \{c, d, f, g\}$, $F = \{a, b, e\}$. Find $D \cup E$, $D \cap E$, and $E \cap F$.

**Solutions:**    $D \cup E = \{a, b, c, d, e, f, g\}$; $D \cap E = \{c, d, f\}$; $E \cap F = \phi$. Sets with no common elements are called **disjoint**.

We say $A$ is a subset of $B$, written $A \subseteq B$ if every element in $A$ is also in $B$. If a set has $n$ elements, it has $2^n$ subsets.

**Example 14:** Write all the subsets of $G = \{a, b, c\}$.

**Solution:**    Because $G$ has 3 elements, there are $2^3 = 8$ subsets. They are: $\phi$, $\{a\}$, $\{b\}$, $\{c\}$, $\{a, b\}$, $\{a, c\}$, $\{b, c\}$, $\{a, b, c\}$.

Two other terms relating to sets that we need to know are:

- The **universe**, which is the set of everything we are talking about.

The universe could be $U = \{$all animals$\}$, or it could be $A = \{$all mammals.$\}$ It depends on what our topic of interest is. Note that $A \subseteq U$.

- A **complement** is the set of elements in the universe not in a specific set. The complement is denoted by a superscript letter $c$, so $A^c$ is the complement of $A$. There are many other symbols in math for the complement.

For example, if $U = \{$all animals$\}$ and $M = \{$all mammals$\}$, then $M^c = \{$reptiles, insects, fish, etc.$\}$

**Example 15:** If the alphabet is the universe, and $A = \{$vowels$\}$, what is $A^c$?

**Solution:**    $A^c = \{$consonants$\}$.

 *If we change the universe, we also automatically change the complement. This type of complement is spelled with two e's; not as "compliment," which is flattering.*

## MATRICES

A matrix is an array of numbers. We will use letters to stand for numbers. $\begin{bmatrix} a & b & c \\ d & e & f \end{bmatrix}$ is a 2 by 3 matrix, written $2 \times 3$. It has 2 rows: row 1, $\begin{bmatrix} a & b & c \end{bmatrix}$; and row 2, $\begin{bmatrix} d & e & f \end{bmatrix}$. It has 3 columns: column 1, $\begin{bmatrix} a \\ d \end{bmatrix}$; column 2, $\begin{bmatrix} b \\ e \end{bmatrix}$; and column 3, $\begin{bmatrix} c \\ f \end{bmatrix}$.

To add two matrices, they must have the same number of rows and columns. You then add the corresponding entries.

**Example 16:** $\begin{bmatrix} a & b & c \\ d & e & f \end{bmatrix} + \begin{bmatrix} r & s & t \\ u & v & w \end{bmatrix} =$

**Solution:**    $\begin{bmatrix} a+r & b+s & c+t \\ d+u & e+v & f+w \end{bmatrix}$

To multiply two matrices, the number of columns of the first matrix must be the same as the number of rows of the second. If we multiplied a 2 × 5 matrix by a 5 × 3 matrix, we would get a 2 × 3 matrix. However, multiplying a 5 × 3 matrix by a 2 × 5 matrix is impossible because 3 (columns in the first matrix) is not equal to 2 (rows in the second matrix).

The next example shows the procedure for multiplying matrices.

**Example 17:** Let $A = \begin{bmatrix} b & c & d \\ f & g & h \end{bmatrix}$ and $B = \begin{bmatrix} j & m \\ k & n \\ l & p \end{bmatrix}$. Find $AB$ and $BA$.

**Solutions:** $A$ is a 2 × 3 matrix, and $B$ is a 3 × 2 matrix. $AB$ would be a 2 × 2 matrix.

The first entry is found by multiplying the first row of $A$ by the first column of $B$. The entry would be the element in the first row, first column of $AB$. We get $bj + ck + dl$. The first row of $A$ multiplied by the second column of $B$ gives the first row, second column of $AB$. We get $bm + cn + dp$. The second row, first column of $AB$ is $fj + gk + hl$, and the second column, second row of $AB$ is $fm + gn + hp$. The matrix $AB$ would look like this:

$$AB = \begin{bmatrix} bj + ck + dl & bm + cn + dp \\ fj + gk + hl & fm + gn + hp \end{bmatrix}$$

Similarly, to get $BA$, we are multiplying a 3 × 2 matrix by a 2 × 3 matrix, so we get a 3 × 3 matrix:

$$BA = \begin{bmatrix} jb + mf & jc + mg & jd + mh \\ kb + nf & kc + ng & kd + nh \\ lb + pf & lc + pg & ld + ph \end{bmatrix}$$

## DETERMINANTS

If we have a square matrix (the number of rows equals the number columns), we can give the matrix a value.

A 2–by–2 determinant, written 2 × 2, looks like this: $\begin{vmatrix} a & b \\ c & d \end{vmatrix}$. Its value is $(ad - bc)$.

**Example 18:** $\begin{vmatrix} 4 & 5 \\ 6 & -7 \end{vmatrix} =$

**Solution:** $4(-7) - 5(6) = -58$.

**Note** *The value of a determinant can be positive, negative, or zero.*

Determinants can be used to solve two equations in two unknowns by the following method.

Suppose we have    $ax + by = m$

$cx + dy = n$

To solve for $x$ and $y$, we would use the following determinants:

$$x = \frac{\begin{vmatrix} m & b \\ n & d \end{vmatrix}}{\begin{vmatrix} a & b \\ c & d \end{vmatrix}} \text{ and } y = \frac{\begin{vmatrix} a & m \\ c & n \end{vmatrix}}{\begin{vmatrix} a & b \\ c & d \end{vmatrix}}.$$

The denominators are the coefficients of the $x$ and $y$ terms. If you are solving for $x$, replace the coefficients of the $x$ numbers in the numerator by $m$ and $n$ from the answer column. Likewise, when you solve for $y$, the $m$ and $n$ replace the coefficients of the $y$ numbers.

> **Example 19:**  Solve for $x$ and $y$ by using determinants:
>
> $$5x - 7y = 29$$
> $$11x + 9y = 10$$
>
> **Solution:**    $x = \dfrac{\begin{vmatrix} 29 & -7 \\ 10 & 9 \end{vmatrix}}{\begin{vmatrix} 5 & -7 \\ 11 & 9 \end{vmatrix}} = \dfrac{(29)(9) - (-7)(10)}{(5)(9) - (-7)(11)} = \dfrac{331}{122}$
>
> $y = \dfrac{\begin{vmatrix} 5 & 29 \\ 11 & 10 \end{vmatrix}}{\begin{vmatrix} 5 & -7 \\ 11 & 9 \end{vmatrix}} = \dfrac{(5)(10) - (29)(11)}{(5)(9) - (-7)(11)} = -\dfrac{269}{122}.$

**Note**  *The beauty of this method is that if the numbers are really messy, there is no problem finding the solution. Any other way, the arithmetic would be much worse.*

**Note**  *If the COMPASS asks you to solve two equations in two unknowns, you may do it any way you want since the test is multiple-choice and you don't need to show your work.*

You can similarly use determinants for three equations in three unknowns. The following method shows how to evaluate a $3 \times 3$ determinant. The *COMPASS* does not ask to solve three equations in three unknowns.

To evaluate $\begin{vmatrix} a & b & c \\ d & e & f \\ g & h & i \end{vmatrix}$, rewrite the first two columns as $\begin{vmatrix} a & b & c \\ d & e & f \\ g & h & i \end{vmatrix} \begin{matrix} a & b \\ d & e \\ g & h \end{matrix}$

The value is the product of the diagonals going from the upper left to the lower right minus the product of the diagonals going from the lower left to the upper right.

In symbols, the value is $aei + bfg + cdh - gec - hfa - idb$.

**Example 20:** Evaluate $\begin{vmatrix} 4 & 5 & 0 \\ 2 & -1 & 1 \\ 0 & 3 & -3 \end{vmatrix}$.

**Solution:** $\begin{vmatrix} 4 & 5 & 0 \\ 2 & -1 & 1 \\ 0 & 3 & -3 \end{vmatrix} \begin{matrix} 4 & 5 \\ 2 & -1 \\ 0 & 3 \end{matrix} =$

$(4)(-1)(-3) + 5(1)(0) + 0(2)(3) - 0(-1)(0) - (3)(1)(4) - (-3)(2)(5) = 30.$

Facts about determinants:

- If one row (or column) is all zeroes, the value of the determinant is 0.

- If two rows (or columns) are equal, the value of the determinant is 0.

- If two rows (or columns) are interchanged, the value is multiplied by $-1$.

 *There are two methods of evaluating a 3 × 3 determinant. This method is the easier, but it will not work for 4 × 4, 5 × 5, or higher-order determinants. However, these methods are beyond the scope of the COMPASS test.*

## TRANSLATIONS

Let's look at the graph of $y = x^2$:

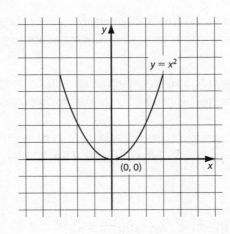

Translations of this graph include the following:

- The same graph 3 units down would be $y = x^2 - 3$.

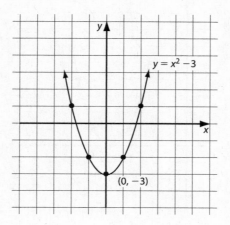

- Or the same graph 5 units up would be $y = x^2 + 5$.

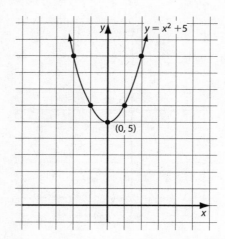

- Or the same graph 4 units to the left would be $y = (x + 4)^2$.

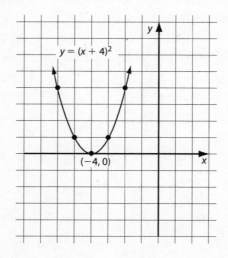

- Or the same graph 4.5 units to the right would be $y = (x - 4.5)^2$.

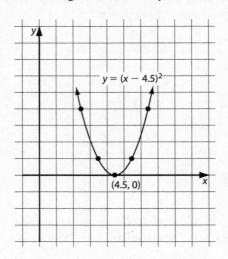

$y = (x - 4.5)^2$

$(4.5, 0)$

- Finally, the same graph upside down would be $y = -x^2$.

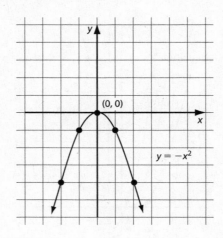

$(0, 0)$

$y = -x^2$

**Example 21:** What would the curve look like if it were 4 units to the left, 3 units down, and upside down?

**Solution:** The equation would be $y = -(x + 4)^2 - 3$.

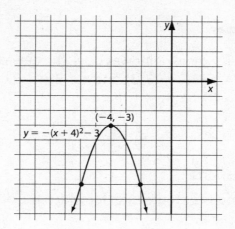

**Note**  *The higher the positive coefficient in front of the $x^2$ term, the quicker the curve goes up. (Conversely, the higher the absolute value of the negative coefficient in front of the $x^2$ term, the quicker the curve goes down.)*

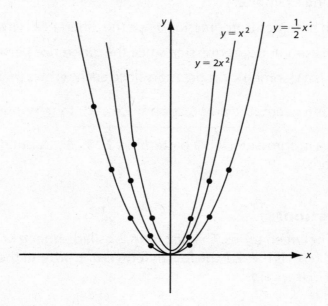

**Note**  *Notice that with trig functions, we did the same thing.*

**Q**  **Let's try a multiple-choice exercise with translations.**

**Exercise 7:**    After moving 4 units to the left and down 6 units, a point on a graph ended up at (3, 3). The original point was

A. (7, 9)                        D. (7, −3)

B. (−1, 9)                     E.  none of these

C. (−1, −3)

**Let's look at the answer.**

**Answer 7:**    **A:** If the original point was $(x, y)$, the new point would be $(x - 4, y - 6)$. So $(x - 4, y - 6) = (3, 3)$, and $x = 7, y = 9$.

# PROGRESSIONS

There are two kinds of progressions:

1.  An **arithmetic progression** is a progression for which the difference between any two neighboring terms is the same.

2.  A **geometric progression** is a progression in which each term is multiplied by the same number to get the next term.

Let's look at the following examples:

- 9, 13, 17, 21… is an arithmetic progression since the difference between terms is 4.

- 4, 3.5, 3, 2.5… is an arithmetic progression since the difference between terms is $-0.5$.

- 5, $-10$, 20, $-40$… is a geometric progression since each term is multiplied by $-2$.

- 486, 162, 54, 18… is a geometric progression since each term is multiplied by $\frac{1}{3}$.

- 1, 4, 9, 16…. is not a progression. It is a pattern: $1^2, 2^2, 3^2, 4^2, \ldots$ but does not fit either of our two progressions.

## Arithmetic Progressions

Let $d =$ the difference between terms. The first term is called either $a$ or $a_1$. The second term is $a_2 = a + d$. The third term is $a + 2d$; the fourth term is $a + 3d$; and the last (or $n$th) term, denoted by $\ell$ or $a_n$ is $a + (n - 1)d$.

The derivation of the sum is adorable, so let's do it.

If we start from the first number, the sum $S = a + (a + d) + (a + 2d) + (a + 3d) + \ldots + [a + (n - 1)d]$.

If we start from the last number, the sum $S = \ell + (\ell - d) + (\ell - 2d) + (\ell - 3d) + \ldots + [\ell - (n - 1)d]$.

Adding these two equations, we see that all the $d$ terms drop out, and we are left with $n$ terms, $2S = n(a + \ell)$.

So $S = \left(\frac{n}{2}\right)(a + \ell)$. Substituting $\ell = a + (n - 1)d$, we get $S = \left(\frac{n}{2}\right)(2a + (n - 1)d)$.

Summary: There are three formulas you need to know for an arithmetic progression:

$\ell = a + (n - 1)d$; $S = \left(\dfrac{n}{2}\right)(a + \ell)$; and $S = \left(\dfrac{n}{2}\right)(2a + (n - 1)d)$.

**Example 22:** Find the 20th term if the first term is 5 and the difference is 3.

**Solution:** $\ell = a + (n - 1)d$; so $\ell = 5 + (20 - 1)3 = 62$.

**Example 23:** Find the sum if the first term is 6, the last term is 7, and the number of terms is 10.

**Solution:** $S = \left(\dfrac{n}{2}\right)(a + \ell)$; $S = \left(\dfrac{10}{2}\right)(6 + 7) = 65$.

**Example 24:** Find the sum if the first term is $-3$, the difference is $-7$, and the number of terms is 8.

**Solution:** $S = \left(\dfrac{n}{2}\right)(2a + (n - 1)d)$; $S = \left(\dfrac{8}{2}\right)[2(-3)+(8-1)(-7)] = -220$.

**Example 25:** An arithmetic sequence is 7, ___, ____, ____, 55. Fill in the missing terms.

**Solution:** We have a 5-term arithmetic series; $a = 7$, and the last term $a_5 = a + 4d = 55$; so $7 + 4d = 55$; $4d = 48$; and $d = 12$. The terms are 7, 7 + 12, 7 + 24, 7 + 36, and 7 + 48, or 7, 19, 31, 43, and 55.

**Example 26:** If the sum is 10, the first term is $-2$, and the difference is 3, find the number of terms.

**Solution:** A toughie! We use the formula $S = \left(\dfrac{n}{2}\right)(2a + (n - 1)d)$; so

$10 = \left(\dfrac{n}{2}\right)[2(-2) + (n - 1)(3)]$. Multiply through by 2, combine terms to get $20 = n(3n - 7)$, or $3n^2 - 7n - 20 = (3n + 5)(n - 4) = 0$. The number of terms must be a positive integer, so $n = 4$, and there are 4 terms.

## Geometric Progressions (Finite)

A geometric progression has a common ratio, $r$. Again, we let $a =$ the first term. The second term is then $ar$; the third term is $ar^2$; the fourth term is $ar^3$; and the last term is $\ell = a_n = ar^{n-1}$.

Again, the proof of the sum is adorable. So let's do it.

We write out the sum. $S = a + ar + ar^2 + ar^3 + \ldots + ar^{n-1}$.

If we multiply by $-r$, we get $-rS = -ar - ar^2 - ar^3 - \ldots - ar^{n-1} - ar^n$.

If we now add these two equations, all of the middle terms add to 0, we get

$S - rS = S(1 - r) = a - ar^n = a(1 - r^n)$.

So $S = \dfrac{a - ar^n}{1 - r}$ or $\dfrac{a(1 - r^n)}{1 - r}$. Substituting $\ell = ar^{n-1}$ in the first expression, we get $S = \dfrac{a - r\ell}{1 - r}$.

In summary,

$\ell = ar^{n-1};\ S = \dfrac{a(1 - r^n)}{1 - r}$ or $S = \dfrac{a - ar^n}{1 - r}$ or $S = \dfrac{a - r\ell}{1 - r}$.

**Note**    *The first form for S is the one that is usually used.*

**Example 27:**    Suppose the first term of a geometric progression is 10 and the ratio is 3; find the fifth term.

**Solution:**    We have $\ell = ar^{n-1}$; so $\ell = (10)(3)^4 = 810$.

**Example 28:**    Find the sum for the geometric progression of Example 27.

**Solution:**    $S = \dfrac{a(1 - r^n)}{1 - r} = \dfrac{10(1 - 3^5)}{1 - 3} = 1{,}210$.

**Example 29:**    We have a 3-term geometric sequence 6, _____, 11. Find the missing term.

**Solution:**    We can set up a proportion here: $\dfrac{6}{x} = \dfrac{x}{11}$; $x^2 = 66$; $x = \pm\sqrt{66}$. These two numbers are sometimes called the **geometric mean**.

## Infinite Geometric Series

Suppose the geometric series goes on forever. Let's see what would happen to the sum. Let's look at the possibilities:

- If $r > 1$: Look at the series 4, 40, 400, 4000,....Clearly, if you added all this up, the sum would be infinite.

- If $r < -1$: Look at the series 10, $-20$, 40, $-80$,....If you stopped after a negative term the sum would be very small (a large negative number). If you stopped after a positive number, the sum would be big. Clearly, there is no one sum.

- If $r = 1$: Look at the series 7, 7, 7, 7,.... The sum of an infinite number of 7's is infinite.

- If $r = -1$: Look at the series 8, $-8$, 8, $-8$, 8, $-8$,.... If you stopped after an odd number, the sum is 8; if you stopped after an even number of terms, the sum is 0. The sum must tend toward one number as $n$ increases, so $r = -1$ is not allowed.

- If $-1 < r < 1$: Let's take $r = \dfrac{1}{10}$, for example. Let's look at $\left(\dfrac{1}{10}\right)^n$ as $n$ gets very big. The sequence becomes $\dfrac{1}{10}, \dfrac{1}{100}, \dfrac{1}{1,000}, \dfrac{1}{10,000},....$ The bigger $n$ gets, the smaller the numbers get. The limit is $r^n = 0$. So the finite geometric sum, $S = \dfrac{a(1-r^n)}{1-r}$, becomes the infinite geometric sum $S = \dfrac{a}{1-r}$.

**Example 30:** Find the infinite sum of 100, 80, 64,....

**Solution:**    We see $a = 100$ and $r = \dfrac{80}{100} = \dfrac{4}{5}$. $S = \dfrac{a}{1-r} = \dfrac{100}{1 - \dfrac{4}{5}} = 500$.

**Example 31:** Change the repeating decimals .3333... and .979797.... to rational numbers.

**Solution:**    We can write .3333... as $.3 + .03 + .003 +...$; $a = .3$ and $r = .1$; so $S = \dfrac{a}{1-r} = \dfrac{.3}{1-.1} = \dfrac{.3}{.9} = \dfrac{1}{3}$. Similarly, we can write .979797... as $.97 + .0097 + .000097 + ...$; so $a = .97$ and $r = .01$; so $S = \dfrac{.97}{1-.01} = \dfrac{97}{99}$.

## SUMMATION

The final topic in this chapter will be summation. The summation symbol is the Greek letter $\Sigma$ (pronounced "sigma"), and it indicates the operation of addition. For computation, the symbol will appear as $\sum\limits_{i=a}^{b}$, where the letter $i$ is the variable (called the index) and $a$ and $b$ represent the lowest and highest values assigned to $i$. Most times, an algebraic expression with the variable $i$ follows the summation symbol. Note that the index may be any letter, but the letters $i, j, k$, and n are most common.

**Example 32:** Find the value of $\sum\limits_{i=1}^{4}(3i+5)$.

**Solution:**    Substitute the values of 1, 2, 3, and 4 for $i$ into the expression $3i + 5$, then add these values. So, $\sum\limits_{i=1}^{4}(3i+5) = [(3)(1) + 5] + [(3)(2) + 5] +$ $[(3)(3) + 5] + [(3)(4) + 5] = 8 + 11 + 14 + 17 = 50$.

**Example 33:** Find the value of $\displaystyle\sum_{n=4}^{6}(n^2+2n)$.

**Solution:** $\displaystyle\sum_{n=4}^{6}(n^2+2n)=[4^2+(2)(4)]+[5^2+(2)(5)]+[6^2+(2)(6)]=[16+8]+$
$[25+10]+[36+12]=107$.

**Example 34:** Find the value of $\displaystyle\sum_{j=2}^{2}(j+10)$.

**Solution:** The lower and upper limits are equal, so we simply substitute 2 for $j$. Then the answer is $2+10=12$.

**Example 35:** Find the value of $\displaystyle\sum_{k=-1}^{2}5$.

**Solution:** At first glance, this appears to be a printing error, but it is not!! Since there is no $k$ in the expression following the summation symbol, this expression is always equal to 5. The fact that $k$ is indexed from $-1$ to 2 tells us that the number 5 must be added for each integer from $-1$ to 2, inclusive. The number of integers from $-1$ to 2 is actually 4. (Don't forget to include zero!) So, the answer is $(4)(5)=20$.

## Chapter 16 Quiz

1. $6!-4!=$
2. $\dfrac{8!}{7!}=$
3. $P(10,3)=$
4. $C(8,2)=$
5. Evaluate $\begin{vmatrix} 0 & 1 & 2 \\ 3 & 4 & 5 \\ -2 & 0 & -1 \end{vmatrix}$.
6. All children must take history or geography. If 30 kids take geography and 40 kids take history, and 27 take both, how many children are there?
7. You have a choice of 5 entrees, 7 desserts, and 10 drinks. How many different meals are there if you can choose one from each group?

For questions 8–13, use the set {4, 6, 7, 8, 9}; an example of a three-digit number is 552.

8. How many three-digit numbers are there?
9. How many three-digit permutations of these numbers are there?

10. How many three-digit odd numbers greater than 700 are there?

11. How many three-digit even numbers greater than 900 are there?

12. How many three-digit numbers have their last two digits the same?

13. How many numbers are less than 800?

14. How many ways can 5 people stand in a line?

15. How many committees of 3 can be formed from 10 people?

For questions 16–20, let the Universe $U = \{a, b, c, d, e, f, g, h\}$; and sets $A = \{a, c, d, f, g\}$; $B = \{a, b, d, f, h\}$; $C = \{c, e, g\}$.

16. Find $A \cup B$.

17. Find $A \cap B$.

18. Find $B \cap C$.

19. Find the complement of $A$.

20. Find the complement of the set of vowels.

21. Find the probability that 3 heads will occur from 3 tosses of a fair coin.

Questions 22 and 23 refer to a jar that has 4 red balls and 5 yellow balls.

22. Find the probability that 2 yellow balls will be picked with replacement.

23. Find the probability that 2 red balls will be picked without replacement.

24. Solve for $x$: $\begin{vmatrix} x & -4 \\ x & x+3 \end{vmatrix} = 8$.

25. If the point $(a, b)$ is translated 4 to the left and 7 up, what are the coordinates of the new point?

26. What is the 20th term in the sequence 100, 94, 88, 82,…?

27. Find the sum of the 20 terms in question 27.

28. Find the 6th term in the sequence 256, 128, 64,…

29. Find the sum of the six terms in question 28.

30. A ball is dropped from 100 feet. It bounces only 80 feet, and keeps this same ratio on all successive bounces. It does this forever. How many feet does the ball travel?

31. Find the geometric means of $-4$ and $-9$.

32. $\displaystyle\sum_{n=0}^{3}(2^n - 3)$

33. $\displaystyle\sum_{i=-5}^{1} 1$

34. $\displaystyle\sum_{j=-2}^{2} j^2$

35. $\displaystyle\sum_{k=-3}^{3} k^3$

36. $\displaystyle\sum_{i=2}^{4} 5^i$

## Answers to Chapter 16 Quiz

1. $720 - 24 = 696$.

2. 8.

3. $10 \times 9 \times 8 = 720$.

4. $\dfrac{8(7)}{1(2)} = 28$.

5. $(0)(4)(-1) + (1)(5)(-2) + 2(3)(0) - (-2)(4)(2) - (-1)(3)(1) - 0(5)(0) = 9$.

6. $30 + 40 - 27 = 43$.

7. $(5)(7)(10) = 350$ different meals.

8. $5(5)(5) = 125$.

9. $P(5, 3) = 5(4)(3) = 60$.

10. $(3)(5)(2) = 30$.

11. $1(5)(3) = 15$.

12. $(5)(5)(1) = 25$. This includes numbers for which all three digits are the same.

13. Read carefully! This includes three-digit numbers, two-digit numbers, and one-digit numbers: $(3)(5)(5) + (5)(5) + 5 = 75 + 25 + 5 = 105$.

14. $5! = 120$.

15. $C(10, 3) = 120$.

16. $\{a, b, c, d, f, g, h\}$.

17. $\{a, d, f\}$.

18. $\phi$.

19. $\{b, e, h\}$.

20. $\{b, c, d, f, g, h\}$.

21. $\left(\dfrac{1}{2}\right)\left(\dfrac{1}{2}\right)\left(\dfrac{1}{2}\right) = \dfrac{1}{8}$.

22. $\left(\dfrac{5}{9}\right)\left(\dfrac{5}{9}\right) = \dfrac{25}{81}$.

23. $\left(\dfrac{4}{9}\right)\left(\dfrac{3}{8}\right) = \dfrac{1}{6}$.

24. $x(x + 3) - (-4)(x) = 8$; $x^2 + 7x - 8$ or $(x + 8)(x - 1) = 0$; $x = -8$ or $x = 1$.

25. $(a - 4, b + 7)$.

26. $a_{20} = a_1 + (n - 1)d$; so $a_{20} = 100 + 19(-6) = -14$.

27. $S = \left(\dfrac{n}{2}\right)(a_1 + a_{20}) = \dfrac{20}{2}(100 + (-14)) = 860$.

28. $a_6 = a_1 r^5 = 256\left(\dfrac{1}{2}\right)^5 = 8$.

29. $S = \dfrac{a(1 - r^n)}{1 - r} = 256\left[\dfrac{1 - \left(\dfrac{1}{2}\right)^6}{1 - \left(\dfrac{1}{2}\right)}\right] = 504$.

30. The first drop gives you 100 feet; then there are two infinite sums (up and down) where the first term is 80 and the ratio is $\dfrac{4}{5} = \dfrac{80}{100}$; so we get $100 + \dfrac{2a}{1 - r} = 100 + \dfrac{2(80)}{1 - \dfrac{4}{5}} =$ 900 feet total distance traveled.

31. $\pm\sqrt{(-4)(-9)} = \pm 6$.

32. 3: $(2^0 - 3) + (2^1 - 3) + (2^2 - 3) + (2^3 - 3) = (1 - 3) + (2 - 3) + (4 - 3) + (8 - 3) = 3$.

33. 7: $1 + 1 + 1 + 1 + 1 + 1 + 1 = 7$.

34. 10: $(-2)^2 + (-1)^2 + 0^2 + 1^2 + 2^2 = 4 + 1 + 0 + 1 + 4 = 10$.

35. 0: The summation appears as $(-3)^3 + (-2)^3 + (-1)^3 + 0^3 + 1^3 + 2^3 + 3^3$. You really do not have to do the actual computations because for any $n$, $n^3 = -(-n)^3$. Then, for example, $3^3 + (-3)^3 = 0$. In a similar way, we can write $2^3 + (-2)^3 = 0$ and $1^3 + (-1)^3 = 0$.

36. 775: $5^2 + 5^3 + 5^4 = 25 + 125 + 625 = 775$.

Now that you have finished the book, let's practice what you've learned.

You should now be ready for the practice tests.

**Answer to "Bob Asks":** $40. The radius is doubled, but $A = \pi r^2$ means the area is $2^2 = 4$ times larger.

# CHAPTER 17: *Practice Test 1*

1. What is the value of $\frac{1}{2}+\left(\frac{3}{4}\times\frac{5}{6}\right)-\left(\frac{3}{4}\div\frac{4}{3}\right)$?

   A. $-\dfrac{83}{72}$

   B. $0$

   C. $\dfrac{1}{8}$

   D. $\dfrac{9}{16}$

   E. $\dfrac{5}{4}$

2. The mean of six numbers is 14. If the smallest number were removed, the mean would become 16. What is the smallest of the six numbers?

   A. 2

   B. 4

   C. 6

   D. 8

   E. 10

3. Twelve is 30% of what number?

   A. 3.6

   B. 4

   C. 36

   D. 40

   E. 360

4.   The foundation of a large house occupies a space of one-fourth of an acre. If the entire property is 4 acres, then on what percent of the property does the house stand?

   A.   4%

   B.   5%

   C.   6.25%

   D.   8.5%

   E.   $16\frac{2}{3}$ %

5.   Sixty percent of the 60 students in a class are girls. How many boys are in this class?

   A.   24

   B.   28

   C.   36

   D.   40

   E.   48

6.   What is the result of rounding off 5.9987 to the nearest tenth?

   A.   6.000

   B.   6.00

   C.   6.0

   D.   6

   E.   5.9

7.   What is the value of $24\frac{6}{7}$ divided by 6?

   A.   $\frac{3}{16}$

   B.   $\frac{9}{16}$

   C.   4

   D.   $4\frac{1}{7}$

   E.   $4\frac{6}{7}$

8. For a certain recipe, if $\frac{1}{3}$ cup of sugar is needed with $\frac{5}{8}$ cup of flour, then how many cups of sugar are needed with 3 cups of flour?

   A. $1\frac{3}{5}$

   B. $2\frac{1}{5}$

   C. $2\frac{1}{2}$

   D. 3

   E. $3\frac{1}{5}$

9. Which of the following is equivalent to $(3x^2y - 4xy) - (5x^2y - 7xy^2)$?
   A. $-210x^2y^2$
   B. $-13x^2y^2$
   C. $-4xy - x^2y$
   D. $-2x^2y - 4xy - 7xy^2$
   E. $-2x^2y - 4xy + 7xy^2$

10. Which of the following is equivalent to $(3x-5)^2$?
    A. $9x^2 - 25$
    B. $9x^2 + 25$
    C. $9x^2 - 30x + 25$
    D. $9x^2 + 30x + 25$
    E. $9x^2 - 15x + 25$

**11.** What are the roots of the equation $x^2 - 7x - 8 = 0$?

   **A.** 1 and 8

   **B.** $-1$ and $-8$

   **C.** $-1$ and 8

   **D.** 1 and $-8$

   **E.** $1, -1, 8,$ and $-8$

**12.** What is the sum of the roots of the equation $5x^2 - 7x - 11 = 0$?

   **A.** $2\frac{1}{5}$

   **B.** $1\frac{2}{5}$

   **C.** $-1\frac{2}{5}$

   **D.** $-2\frac{1}{5}$

   **E.** $-15\frac{2}{5}$

**13.** What is the solution for $x$ in the equation $3(5x - 7) - 4(2x - 5) = 1$?

   **A.** 0

   **B.** $\frac{2}{23}$

   **C.** $\frac{2}{7}$

   **D.** 6

   **E.** $6\frac{15}{23}$

**14.** If $x = -3$ and $y = -10$, what is the value of $xy^2 - (xy)^2$?

   **A.** 600

   **B.** 0

   **C.** $-300$

   **D.** $-600$

   **E.** $-1,200$

**15.** If $x \neq 0, 3$, what is the simplified form of $\dfrac{9 - x^2}{x^2 - 3x}$?

   **A.** $-\dfrac{3}{x}$

   **B.** $\dfrac{3}{x}$

   **C.** $\dfrac{x+3}{x}$

   **D.** $\dfrac{-x-3}{x}$

   **E.** 3

**16.** Which of the following is equivalent to $\dfrac{4}{x^2 - 4} + \dfrac{5}{x^2 - 4x + 4}$?

   **A.** $\dfrac{9}{2x(x-2)}$

   **B.** $\dfrac{9}{(x-2)^3(x+2)}$

   **C.** $\dfrac{9}{(x-2)^2(x+2)}$

   **D.** $\dfrac{9x-2}{(x-2)^2(x+2)}$

   **E.** $\dfrac{9x+2}{(x-2)^2(x+2)}$

**17.** What is the simplified form of $\dfrac{8x^6 - 12x^4 - 4x^2}{4x^2}$?

    **A.**  $8x^6 - 12x^4 - 1$

    **B.**  $8x^4 - 12x^2 - 1$

    **C.**  $4x^4 - 8x^2 - 1$

    **D.**  $2x^4 - 3x^2 - 1$

    **E.**  $2x^4 - 3x^2$

**18.** The equation of line 1 is $3x + 5y = -1$. If line 2 is perpendicular to line 1, which of the following could be the equation of line 2?

    **A.**  $3x + 5y = 1$

    **B.**  $5x - 3y = 23$

    **C.**  $-5x - 3y = 17$

    **D.**  $3x - 5y = 1$

    **E.**  $3x - y = 5$

**19.** What is the distance between the points $(4, -5)$ and $(3, 3)$?

    **A.**  9

    **B.**  $\sqrt{65}$

    **C.**  8

    **D.**  $\sqrt{53}$

    **E.**  $\sqrt{5}$

**20.** What is the value of $125^{\frac{-4}{3}}$?

    **A.**  $-625$

    **B.**  $-166\dfrac{2}{3}$

    **C.**  $-\dfrac{1}{625}$

    **D.**  $\dfrac{1}{625}$

    **E.**  625

21.   What is the simplified form of $\sqrt[3]{32x^{11}}$ ?

A.   $4x^5\sqrt[3]{2x}$

B.   $2x^3\sqrt[3]{4x^2}$

C.   $2x^2\sqrt[3]{2x}$

D.   $8x^9\sqrt[3]{4x^2}$

E.   $16x^{10}\sqrt[3]{2x}$

22.   How many gallons of a liquid that contains 30% alcohol must be mixed with a liquid that contains 60% alcohol in order to produce 12 gallons of a mixture that contains 50% alcohol?

A.   2

B.   3

C.   4

D.   6

E.   8

23.   Which of the following is equivalent to $\dfrac{\sqrt{a}}{\sqrt{a}+\sqrt{b}}$ ?

A.   $\dfrac{a+\sqrt{ab}}{a+b}$

B.   $\dfrac{a-\sqrt{ab}}{a-b}$

C.   $\dfrac{a-\sqrt{b}}{a-b}$

D.   $\dfrac{a+\sqrt{b}}{a+b}$

E.   $\dfrac{1}{\sqrt{b}}$

24.    John can do a job in 6 hours and Mary can do this same job in 4 hours. Working together, how many hours will they need to do this job?

A.    $3\dfrac{1}{2}$

B.    $3\dfrac{1}{4}$

C.    3

D.    $2\dfrac{3}{4}$

E.    $2\dfrac{2}{5}$

25.    The sum of five consecutive integers is 115. What is the sum of the largest two integers?

A.    21

B.    23

C.    45

D.    49

E.    53

26.    What is the slope of a line that is parallel to the graph of $8 + 3y + x = 0$?

A.    3

B.    1

C.    $\dfrac{1}{3}$

D.    $-\dfrac{1}{3}$

E.    $-3$

27. For a certain rectangle, the diagonal is 8 and one side is 5. What is the length of the other side?

    A.   3

    B.   $\sqrt{39}$

    C.   $\sqrt{89}$

    D.   $\sqrt{148}$

    E.   13

28. A box has a length of 4x, a width of 3x, and a height of 5x. What is its surface area?

    A.   $47x^2$

    B.   $60x^2$

    C.   $60x^3$

    D.   $94x^2$

    E.   $188x^2$

29. The angles of a triangle are in the ratio 3:5:7. How many degrees are in the largest angle?

    A.   12

    B.   36

    C.   72

    D.   84

    E.   96

30. Each of the diameter of the base of a cylinder and its height is 20 units. What is the number of cubic units in the volume?

    A.   $\dfrac{1{,}000}{3}\pi$

    B.   $\dfrac{2{,}000}{3}\pi$

    C.   $1{,}000\pi$

    D.   $2{,}000\pi$

    E.   $8{,}000\pi$

**31.** For a certain trapezoid, the two bases are 20 and 40. If the area is 480, what is the height?

   A.   4

   B.   6

   C.   8

   D.   12

   E.   16

**32.** In a 30°-60°-90° right triangle, the hypotenuse is $8\sqrt{3}$. What is the area?

   A.   24

   B.   $24\sqrt{3}$

   C.   48

   D.   $48\sqrt{3}$

   E.   $96\sqrt{3}$

**33.** In an isosceles triangle, the measure of one base angle is 27°. How many degrees are in the exterior angle at the vertex?

   A.   27

   B.   54

   C.   90

   D.   114

   E.   126

**34.** The circumference of a circle is 16π What is the area?

   A.   8π

   B.   16π

   C.   48π

   D.   64π

   E.   256π

35. The first term of an arithmetic sequence is $b$ and the difference between successive terms is 6. What is the 20th term?

 A. $b + 114$

 B. $b + 120$

 C. $b + 126$

 D. $b + 132$

 E. $b + 140$

36. If $a + bi = 4i(3 + i)$, what is the value of $ab$?

 A. $-48$

 B. $-8$

 C. 8

 D. 32

 E. 48

37. What is the value of $\frac{10!}{(8!)(2!)}$?

 A. 45

 B. 60

 C. 75

 D. 90

 E. 180

38. What is the value of $\sum_{n=1}^{5}(\frac{1}{n} - \frac{1}{n+1})$?

 A. $\frac{2}{3}$

 B. $\frac{3}{4}$

 C. $\frac{5}{6}$

 D. $\frac{9}{10}$

 E. 1

**39.** What is the value of $\log_{16}32$ ?

    A.   1

    B.   $1\dfrac{1}{4}$

    C.   $1\dfrac{1}{2}$

    D.   $1\dfrac{3}{4}$

    E.   2

**40.** What are the coordinates of the vertex of the graph of $f(x) = 2x^2 - 8x + 100$?

    A.   $(-2, 124)$

    B.   $(2, 92)$

    C.   $(4, 100)$

    D.   $(6, 124)$

    E.   $(10, 220)$

**41.** What is the value of $x$ in the equation $\log_{125} x = -\dfrac{2}{3}$?

    A.   5

    B.   $\dfrac{1}{5}$

    C.   $\dfrac{1}{25}$

    D.   $-\dfrac{1}{5}$

    E.   $-5$

**42.** The expression $3\log_2 x - 4\log_2 y + 5\log_2 z$ can be simplified to which one of the following?

   A.   $\log_2 \dfrac{15xz}{4y}$

   B.   $\log_2 \dfrac{x^3}{y^4 z^5}$

   C.   $\log_2 \dfrac{x^3 z^5}{y^4}$

   D.   $12\log_2 xyz$

   E.   $60\log_2 \dfrac{xz}{y}$

**43.** If $\tan A = \dfrac{5}{12}$ and $0° < A < 90°$, what is the value of $\sin A$?

   A.   $\dfrac{5}{13}$

   B.   $\dfrac{12}{13}$

   C.   $\dfrac{13}{12}$

   D.   $\dfrac{12}{5}$

   E.   $\dfrac{13}{5}$

**44.** If $\cos(5x) = 0$, what is the minimum positive value of $x$ in degrees?

   A.   9

   B.   18

   C.   36

   D.   45

   E.   90

**45.** Which one of the following is equivalent to $\sin^2 x + \cos^2 x + \cot^2 x$?

  A.  $\csc^2 x$

  B.  $\sec^2 x$

  C.  $\cot^2 x$

  D.  $\tan^2 x$

  E.  $\sin^2 x$

**46.** A viewer is looking up at an angle of 63° to the top of a building. If the viewer is 100 feet from the base of the building, which of the following represents the height of the building? (Ignore the height of the viewer.)

  A.  100 tan 63°

  B.  $\dfrac{100}{\tan 63°}$

  C.  100 sin 63°

  D.  $\dfrac{100}{\sin 63°}$

  E.  100 cos 63°

**47.** If $\cos A = \dfrac{1}{2}$ and $0° < A < 90°$, what is the value of $\sin(2A)$?

  A.  $-\dfrac{\sqrt{3}}{2}$

  B.  $-\dfrac{\sqrt{2}}{2}$

  C.  $\dfrac{1}{2}$

  D.  $\dfrac{\sqrt{2}}{2}$

  E.  $\dfrac{\sqrt{3}}{2}$

**48.** What is the value of $\sin(\cos^{-1}\dfrac{7}{25})$ ?

    A.   $\dfrac{25}{7}$

    B.   $\dfrac{24}{7}$

    C.   $\dfrac{25}{24}$

    D.   $\dfrac{24}{25}$

    E.   $\dfrac{7}{24}$

**49.** What is the solution for $x$ in the equation $4^{2x+3} = 32$?

    A.   $-\dfrac{1}{2}$

    B.   $-\dfrac{1}{4}$

    C.   $\dfrac{1}{4}$

    D.   $\dfrac{1}{2}$

    E.   $\dfrac{5}{2}$

**50.** What is the simplified expression for $i^{79}$?

    A.   $-1$

    B.   $0$

    C.   $1$

    D.   $i$

    E.   $-i$

## SOLUTIONS

1.    **D:** $\frac{1}{2}+\left(\frac{3}{4}\times\frac{5}{6}\right)-\left(\frac{3}{4}\div\frac{4}{3}\right)=\frac{1}{2}+\frac{5}{8}-\frac{9}{16}=\frac{8}{16}+\frac{10}{16}-\frac{9}{16}=\frac{9}{16}.$

2.    **B:** The total of the six numbers is $(6)(14) = 84$. If one number were removed, the total of the five remaining numbers would be $(5)(16) = 80$. The removed number is $84 - 80 = 4$.

3.    **D:** $12 \div 0.30 = 40$.

4.    **C:** $\frac{1}{4}\div 4=\frac{1}{16}=6.25\%.$

5.    **A:** The number of girls is $(0.60)(60) = 36$, so there are $60 - 36 = 24$ boys.

6.    **C:** The digit in the hundredths place (9) is greater than 5. Drop this digit and all other digits to its right and increase the tenths digit (9) by one unit. This means that the tenths digit becomes 0, and this forces the units digit to increase by 1. The result is 6.0.

7.    **D:** $24\frac{6}{7}\div 6=\frac{174}{7}\times\frac{1}{6}=\frac{174}{42}=\frac{29}{7}=4\frac{1}{7}.$

8.    **A:** Let $x$ represent the number of cups of sugar. Then $\frac{\frac{1}{3}}{\frac{5}{8}}=\frac{x}{3}$. Cross-multiply to get $\frac{5}{8}x=1$. Thus, $x = 1\div\frac{5}{8}=\frac{8}{5}=1\frac{3}{5}$.

9.    **E:** Removing the parentheses from the original expression, we get $3x^2y - 4xy - 5x^2y + 7xy^2$. The first and third terms can be combined as $-2x^2y$, so that the answer becomes $-2x^2y - 4xy + 7xy^2$.

10.    **C:** $(3x - 5)^2 = 9x^2 - 15x - 15x + 25 = 9x^2 - 30x + 25$.

11.    **C:** Using factoring, $x^2 - 7x - 8 = 0$ can be written as $(x + 1)(x - 8) = 0$. Then $x + 1 = 0$ or $x - 8 = 0$, which leads to the roots $-1$ or $8$.

12.    **B:** For the equation $ax^2 + bx + c = 0$, the sum of the roots is $\frac{-b}{a}$. Thus, the sum of the roots of $5x^2 - 7x - 11 = 0$ is $\frac{-(-7)}{5}=\frac{7}{5}=1\frac{2}{5}$.

13.    **C:** $3(5x - 7) - 4(2x - 5) = 1$ becomes $15x - 21 - 8x + 20 = 1$. Combine like terms to get $7x - 1 = 1$. Then $7x = 2$, so $x = \frac{2}{7}$.

**14.**   **E:** $(-3)(-10)^2 - [(-3)(-10)]^2 = (-3)(100) - (30)^2 = -300 - 900 = -1,200.$

**15.**   **D:** $\dfrac{9-x^2}{x^2-3x} = \dfrac{(3-x)(3+x)}{x(x-3)} = -\dfrac{3+x}{x} = \dfrac{-x-3}{x}.$

**16.**   **E:** $\dfrac{4}{x^2-4} + \dfrac{5}{x^2-4x+4} = \dfrac{4}{(x+2)(x-2)} + \dfrac{5}{(x-2)^2} =$

$\dfrac{4(x-2)}{(x+2)(x-2)^2} + \dfrac{5(x+2)}{(x+2)(x-2)^2} = \dfrac{4x-8+5x+10}{(x+2)(x-2)^2} = \dfrac{9x+2}{(x-2)^2(x+2)}.$

**17.**   **D:** $\dfrac{8x^6-12x^4-4x^2}{4x^2} = \dfrac{8x^6}{4x^2} - \dfrac{12x^4}{4x^2} - \dfrac{4x^2}{4x^2} = 2x^4 - 3x^2 - 1.$

**18.**   **B:** $3x + 5y = -1$ can be written as $y = -\dfrac{3}{5}x - \dfrac{1}{5}$, which means that the slope of line 1 is $-\dfrac{3}{5}$. Then the slope of line 2 must be the negative reciprocal of $-\dfrac{3}{5}$, which is $\dfrac{5}{3}$. The equation $5x - 3y = 23$ can be written as $y = \dfrac{5}{3}x - \dfrac{23}{3}$, for which the slope is $\dfrac{5}{3}$.

The slopes of the lines corresponding to answer choices A, C, D, and E are $-\dfrac{3}{5}, -\dfrac{5}{3}$, $\dfrac{3}{5}$, and 3, respectively.

**19.**   **B:** The distance is $\sqrt{(3-4)^2 + (3-[-5])^2} = \sqrt{(-1)^2 + 8^2} = \sqrt{65}.$

**20.**   **D:** $125^{-\frac{4}{3}} = \dfrac{1}{125^{\frac{4}{3}}} = \dfrac{1}{(\sqrt[3]{125})^4} = \dfrac{1}{5^4} = \dfrac{1}{625}.$

**21.**   **B:** $\sqrt[3]{32x^{11}} = \sqrt[3]{(2\times2\times2)\times(2\times2)\times(xxx)(xxx)(xxx)\times(xx)} = 2x^3\sqrt[3]{4x^2}.$

**22.**   **C:** Let $x$ represent the required number of gallons. Then $0.30x + 0.60(12 - x) = (0.50)(12)$, which becomes $0.30x + 7.2 - 0.60x = 6.$ Then $-0.30x = -1.2$, so $x = 4.$

**23.**   **B:** $\dfrac{\sqrt{a}}{\sqrt{a}+\sqrt{b}} = \left(\dfrac{\sqrt{a}}{\sqrt{a}+\sqrt{b}}\right)\left(\dfrac{\sqrt{a}-\sqrt{b}}{\sqrt{a}-\sqrt{b}}\right) = \dfrac{a-\sqrt{ab}}{a-\sqrt{ab}+\sqrt{ab}-b} = \dfrac{a-\sqrt{ab}}{a-b}.$

**24.**   **E:** Let $x$ represent the required number of hours. Then $\dfrac{x}{6} + \dfrac{x}{4} = 1.$ Multiply the equation by 12 to get $2x + 3x = 12.$ Then $5x = 12$, so $x = 2\dfrac{2}{5}.$

**25.**   **D:** The middle (third) number must be $\dfrac{115}{5} = 23$. Then the two largest numbers are 24 and 25, and their sum is 49.

**26.**   **D:** Rewrite the original equation as $3y = -x - 8$, which becomes $y = -\dfrac{1}{3}x - \dfrac{8}{3}$. Its graph is a line with a slope of $-\dfrac{1}{3}$, which must match the slope of any parallel line.

**27.**   **B:** Let $x$ represent the length of the other side. By the Pythagorean theorem, $x^2 + 5^2 = 8^2$, so $x^2 = 8^2 - 5^2 = 39$. Thus, $x = \sqrt{39}$.

**28.**   **D:** The surface area is $(2)(4x)(3x) + (2)(3x)(5x) + (2)(4x)(5x) = 24x^2 + 30x^2 + 40x^2 = 94x^2$.

**29.**   **D:** The three angles can be represented as $3x$, $5x$, and $7x$. Then $3x + 5x + 7x = 180$. So, $15x = 180$, which means that $x = 12$. Thus, the number of degrees in the largest angle is $(7)(12) = 84$.

**30.**   **D:** The radius is 10, so the volume in cubic units is $(\pi)(10^2)(20) = 2{,}000\pi$.

**31.**   **E:** Let $h$ represent the height. Then $\dfrac{1}{2}h(20 + 40) = 480$. This equation simplifies to $30h = 480$, so $h = 16$.

**32.**   **B:** The shorter leg is one-half the hypotenuse, which is $4\sqrt{3}$. The longer leg is the product of the shorter leg and $\sqrt{3}$, which is $4\sqrt{3} \times \sqrt{3} = 12$. Thus, the area of the triangle is $\left(\dfrac{1}{2}\right)(4\sqrt{3})(12) = 24\sqrt{3}$.

**33.**   **B:** The measure of each base angle must be 27°. The number of degrees in the vertex angle is $180 - 27 - 27 = 126$. Thus, the number of degrees in the exterior angle at the vertex is $180 - 126 = 54$.

**34.**   **D:** The radius is $\dfrac{16\pi}{2\pi} = 8$, so the area is $(\pi)(8^2) = 64\pi$.

**35.**   **A:** The $n^{\text{th}}$ term of an arithmetic sequence is represented by $a + (n - 1)(d)$, where $a$ is the first term, $n$ is the specific term, and $d$ is the common difference between successive terms. Then the 20th term of the sequence with a first term of $b$ and a common difference of 6 is $b + (19)(6) = b + 114$.

**36.**   **A:** $4i(i + 3) = 4i^2 + 12i = -4 + 12i$. Thus, $ab = (-4)(12) = -48$.

**37.** **A:** $\dfrac{10!}{(8!)(2!)} = \dfrac{(10)(9)(8)(\cdots)(2)(1)}{[(8)(7)(6)(\cdots)(2)(1)][(2)(1)]}$. By canceling $(8)(7)(6)$ ( ) $(2)$ $(1)$ from both

numerator and denominator, we get $\dfrac{(10)(9)}{(2)(1)} = 45$.

**38.** **C:** $\displaystyle\sum_{n=1}^{5}(\dfrac{1}{n} - \dfrac{1}{n+1}) = (1 - \dfrac{1}{2}) + (\dfrac{1}{2} - \dfrac{1}{3}) + (\dfrac{1}{3} - \dfrac{1}{4}) + (\dfrac{1}{4} - \dfrac{1}{5}) + (\dfrac{1}{5} - \dfrac{1}{6}) =$

$1 + (-\dfrac{1}{2} + \dfrac{1}{2}) + (-\dfrac{1}{3} + \dfrac{1}{3}) + (-\dfrac{1}{4} + \dfrac{1}{4}) + (-\dfrac{1}{5} + \dfrac{1}{5}) - \dfrac{1}{6}$. By grouping the numbers in this way,

we notice that the expression reduces to $1 - \dfrac{1}{6} = \dfrac{5}{6}$. (This is called a telescoping series.)

**39.** **B:** Let $x = \log_{16}32$. Then $16^x = 32$. By changing both bases to a base of 2, we get

$(2^4)^x = 2^5$. So $4x = 5$, which means that $x = 1\dfrac{1}{4}$.

**40.** **B:** The $x$-coordinate of the vertex for the graph of $f(x) = ax^2 + bx + c$ is $-\dfrac{b}{2a}$. So, for

$f(x) = 2x^2 - 8x + 100$, the $x$-coordinate of the vertex is $-\dfrac{-8}{(2)(2)} = 2$. The correspond-

ing $y$-coordinate can be found by direct substitution. $(2)(2)^2 - (8)(2) + 100 = 92$.

Thus, the vertex is located at $(2, 92)$.

**41.** **C:** The equation $\log_{125} x = -\dfrac{2}{3}$ is equivalent to $x = 125^{-\frac{2}{3}}$. Rewrite $125^{-\frac{2}{3}}$ as

$\dfrac{1}{125^{\frac{2}{3}}} = \dfrac{1}{\left(\sqrt[3]{125}\right)^2} = \dfrac{1}{5^2} = \dfrac{1}{25}$.

**42.** **C:** Using the exponential rules of logarithms, $3\log_2 x - 4\log_2 y + 5\log_2 z = \log_2 x^3 -$

$\log_2 y^4 + \log_2 z^5$. Now using the multiplication and division rules of logarithms,

$\log_2 x^3 - \log_2 y^4 + \log_2 z^5 = \log_2 \dfrac{x^3 z^5}{y^4}$.

**43.** **A:** Using a right triangle, the tangent ratio equals the opposite side over the adjacent

side. The hypotenuse equals $\sqrt{5^2 + 12^2} = \sqrt{25 + 144} = \sqrt{169} = 13$. The sine ratio

equals the opposite side over the hypotenuse, which is $\dfrac{5}{13}$.

**44.** **B:** The inverse cosine of 0 is 90º, which means that $\cos 90° = 0$. Thus, $5x = 90$,

so $x = 18$.

**45.**    **A:** Use the two trigonometric identities (a) $\sin^2 x + \cos^2 x = 1$ and (b) $1 + \cot^2 x = \csc^2 x$.

**46.**    **A:** Let $x$ represent the height of the building. Then $\tan 63° = \dfrac{x}{100}$, which is equivalent to $x = 100 \tan 63°$.

**47.**    **E:** $A = \cos^{-1}\left(\dfrac{1}{2}\right) = 60°$. Then $2A = 120°$ and $\sin 120° = \dfrac{\sqrt{3}}{2}$.

**48.**    **D:** Let A represent the angle in a right triangle for which $\cos A = \dfrac{7}{25}$. The adjacent side is 7 and the hypotenuse is 25, so the opposite side is $\sqrt{25^2 - 7^2} = \sqrt{625 - 49} = \sqrt{576} = 24$. Thus, $\sin A = \dfrac{24}{25}$.

**49.**    **B:** Rewrite the equation in base 2. Then $(2^2)^{2x+3} = 2^5$, and equating exponents, $4x + 6 = 5$. This means that $4x = 1$; thus $x = -\dfrac{1}{4}$.

**50.**    **E:** The powers of $i$ are in a cycle of 4, so that $i^1 = i, i^2 = -1, i^3 = -i$, and $i^4 = 1$. When 79 is divided by 4, the remainder is 3. Therefore, $i^{79} = i^3 = -i$.

# COMPASS – PRACTICE TEST 1 ANSWER KEY

| Question Number | Answer Key | Content Domain | Question Number | Answer Key | Content Domain |
|---|---|---|---|---|---|
| 1 | D | Pre-Algebra | 26 | D | Algebra |
| 2 | B | Pre-Algebra | 27 | B | Geometry |
| 3 | D | Pre-Algebra | 28 | D | Geometry |
| 4 | C | Pre-Algebra | 29 | D | Geometry |
| 5 | A | Pre-Algebra | 30 | D | Geometry |
| 6 | C | Pre-Algebra | 31 | E | Geometry |
| 7 | D | Pre-Algebra | 32 | B | Geometry |
| 8 | A | Pre-Algebra | 33 | B | Geometry |
| 9 | E | Algebra | 34 | D | Geometry |
| 10 | C | Algebra | 35 | A | College Algebra |
| 11 | C | Algebra | 36 | A | College Algebra |
| 12 | B | Algebra | 37 | A | College Algebra |
| 13 | C | Algebra | 38 | C | College Algebra |
| 14 | E | Algebra | 39 | B | College Algebra |
| 15 | D | Algebra | 40 | B | College Algebra |
| 16 | E | Algebra | 41 | C | College Algebra |
| 17 | D | Algebra | 42 | C | College Algebra |
| 18 | B | Algebra | 43 | A | Trigonometry |
| 19 | B | Algebra | 44 | B | Trigonometry |
| 20 | D | College Algebra | 45 | A | Trigonometry |
| 21 | B | College Algebra | 46 | A | Trigonometry |
| 22 | C | Algebra | 47 | E | Trigonometry |
| 23 | B | Algebra | 48 | D | Trigonometry |
| 24 | E | Algebra | 49 | B | College Algebra |
| 25 | D | Algebra | 50 | E | College Algebra |

# CHAPTER 18: *Practice Test 2*

1. What is the value of $4 - 7 - (-7)^2$?

   A. 46

   B. 38

   C. $-17$

   D. $-29$

   E. $-52$

2. In a batch of 800 light bulbs, four of them are faulty. What percent of the light bulbs is faulty?

   A. 50

   B. 5

   C. 0.5

   D. 0.05

   E. 0.005

3. What is the value of $\sqrt{\dfrac{8}{18}}$ ?

   A. 1.5

   B. 1

   C. $0.\overline{6}$

   D. $0.\overline{4}$

   E. $0.\overline{3}$

4. What is the solution of $x$ in the equation $\dfrac{x-7}{3} = \dfrac{x-3}{7}$ ?

   A. 0

   B. $\dfrac{7}{3}$

   C. 3

   D. 7

   E. 10

5. Consider the following ten grades: 100, 100, 100, 100, 100, 99, 99, 98, 98, 86. What is the sum of the mean, median, and mode?

   A. 297

   B. 297.5

   C. 298

   D. 298.5

   E. 299

6. In scientific notation, what is the sum of 7,000,000 and 42,500?

   A. $7.425 \times 10^{10}$

   B. $7.425 \times 10^{5}$

   C. $7.425 \times 10^{6}$

   D. $7.0425 \times 10^{6}$

   E. $7.0425 \times 10^{5}$

7. 300 is 15% of what number?

   A. 45

   B. 90

   C. 200

   D. 2,000

   E. 3,000

8. Which of the following is equivalent to $6^{10} \times 6^{22}$?

   A. $36^{220}$

   B. $36^{32}$

   C. $6^{220}$

   D. $6^{32}$

   E. $\dfrac{1}{6^{12}}$

9.  If 12 children's movie tickets cost $4 each and 8 adult movie tickets cost $10 each, what is the mean cost of all these tickets?

    A.  $6.00

    B.  $6.40

    C.  $6.60

    D.  $6.80

    E.  $7.00

10. What are the roots of $x^3 + 4x^2 - 12x = 0$ ?

    A.  $0, 2, 6$

    B.  $1, 2, 6$

    C.  $0, -2, -6.$

    D.  $0, -2, 6$

    E.  $0, 2, -6$

11. What is the simplified expression for $10a - 5b - (5b - 2c)$?

    A.  $10a - 2c$

    B.  $10a + 2c$

    C.  $10a - 10b + 2c.$

    D.  $10a + 25b + 2c.$

    E.  $10a - 10b - 2c$

12. What is the simplified expression for $\sqrt{50x^5} + 5x\sqrt{8x^3}$ ?

    A.  $15x^2\sqrt{2x}$

    B.  $100x^5$

    C.  $40x^2\sqrt{2x}$

    D.  $x^4\sqrt{58}$

    E.  $24x\sqrt{8x}$

13.   What is the solution for $x$ in the equation $4x + 5 + 6 + 7x = 0$?

A.   $-1$

B.   $0$

C.   $1$

D.   $11$

E.   $121$

14.   $X$ years ago, Tom was $Y$ years old.  How old will Tom be in $Z$ years?

A.   $X + Y + Z$

B.   $X + Y - Z$

C.   $X - Y + Z$

D.   $Y - X + Z$

E.   $Y - X - Z$

15.   Two less than three times a number is the same as four more than five times that number. If $x$ represents the number, which of the following equations is correct?

A.   $2 - 3x = 4 + 5x$

B.   $2x - 3 = 5x + 4$

C.   $3x - 2 = 5x + 4$

D.   $2 - 3x = 4x + 5$

E.   $2x - 3 = 5 + 4x$

16.   Which of the following is equivalent to $(3ab^2c^3)^2$?

A.   $6ab^4c^5$

B.   $6ab^4c^6$

C.   $9ab^4c^5$

D.   $9a^2b^4c^6$

E.   $27a^2b^4c^6$

**17.** What is (are) the solution(s) for $x$ in the equation $|x - 5| = |x + 5|$?

    **A.** No solution

    **B.** 5

    **C.** −5

    **D.** 5 and −5

    **E.** 0

**18.** If $x \neq 1,5$, what is the reduced form of $\dfrac{x^2 - 4x + 3}{x^2 - 6x + 5}$?

    **A.** $\dfrac{-2x + 3}{-6x + 5}$

    **B.** $-\dfrac{2}{5}$

    **C.** $\dfrac{x - 3}{x - 5}$

    **D.** $\dfrac{3}{5}$

    **E.** $\dfrac{2}{3}x + \dfrac{3}{5}$

**19.** Which of the following is equivalent to $\dfrac{(10a^4)^4(2a^6)^3}{16a^{20}}$?

    **A.** $5{,}000a^{14}$

    **B.** $\dfrac{1}{5a^3}$

    **C.** $5a^{14}$

    **D.** $\dfrac{5{,}000}{a^3}$

    **E.** $5\,a^{268}$

**20.**   What is the simplified form of $\dfrac{10}{9ab^2}+\dfrac{5}{6b}$ ?

   A.   $\dfrac{5}{18ab^3}$

   B.   $\dfrac{5}{18ab^2}$

   C.   $\dfrac{5}{6ab^2}$

   D.   $\dfrac{20+15ab}{18ab^2}$

   E.   $\dfrac{35}{18b}$

**21.**   What is the expanded form of $(4x^2 - 5y^3)^2$?

   A.   $16x^4 + 25y^9$

   B.   $16x^4 + 25y^6$

   C.   $16x^4 - 20x^2y^3 + 25y^6$

   D.   $16x^4 - 40x^2y^3 + 25y^6$

   E.   $16x^4 - 25y^6$

**22.**   What is the slope of a line that is perpendicular to the graph of $3x + 4y = 5$?

   A.   $\dfrac{4}{3}$

   B.   $\dfrac{3}{4}$

   C.   $-\dfrac{3}{4}$

   D.   $-\dfrac{4}{5}$

   E.   $-\dfrac{4}{3}$

23. What is the value of $4^{-\frac{3}{2}} - \dfrac{1}{125^{-\frac{1}{3}}}$ ?

    A. $-5$

    B. $-\dfrac{39}{8}$

    C. $-\dfrac{5}{8}$

    D. $\dfrac{5}{8}$

    E. $\dfrac{8}{5}$

24. What is the distance between the points $(\sqrt{2}, \sqrt{2}-1)$ and $(\sqrt{2}+1, \sqrt{2})$?

    A. $\sqrt{2}-2$

    B. $0$

    C. $1$

    D. $\sqrt{2}$

    E. $\sqrt{2}+2$

25. Using only positive exponents, which of the following is equivalent to $\dfrac{a^{-4}b^{-5}c^{8}}{a^{6}b^{-2}c^{-3}}$ ?

    A. $a^{10}b^{7}c^{11}$

    B. $\dfrac{1}{a^{10}b^{7}c^{11}}$

    C. $\dfrac{c^{11}}{a^{10}b^{3}}$

    D. $\dfrac{a^{10}b^{3}}{c^{11}}$

    E. $\dfrac{c^{512}}{a^{24}b^{32}}$

**26.** What is the simplified form of $\dfrac{(x^4+x^2)^2}{x^4}$?

   **A.** $x^4 + 2x^2 + 1$

   **B.** $x^4 + 2x^2$

   **C.** $x^4 + 1$

   **D.** $x^{12}$

   **E.** $x^8$

**27.** What is an equivalent form of $\sqrt[3]{b^2} \times \sqrt[5]{b^7}$?

   **A.** $\sqrt[15]{b^{31}}$

   **B.** $\sqrt[15]{b^{14}}$

   **C.** $\sqrt[243]{b^{128}}$

   **D.** $\sqrt[8]{b^9}$

   **E.** Cannot be simplified

**28.** At 12:00 noon, a train left New York City going west.  At 2:00 PM, another train left New York City going west on a parallel track.  The second train traveled 40 miles per hour faster than the first train. At 6:00 PM, the second train overtook the first train.  How many miles from New York City did the second train overtake the first train?

   **A.** 80

   **B.** 120

   **C.** 240

   **D.** 360

   **E.** 480

**29.** Which of the following is equivalent to $(2 + 3i)(3 + i)$?

    **A.** $3 - 11i$

    **B.** $-3 + 11i$

    **C.** $11 + 3i$

    **D.** $3 + 11i$

    **E.** $11 - 3i$

30. How many terms are contained in the arithmetic series $9 + 11 + 13 + \ldots + 243$?

    **A.** 117

    **B.** 118

    **C.** 119

    **D.** 120

    **E.** 121

**31.** If the determinant $\begin{vmatrix} x & 4 \\ 6 & x-2 \end{vmatrix} = 11$, what are the values of $x$?

    **A.** 4 or $-6$

    **B.** $-4$ or 6

    **C.** $-7$ or 5

    **D.** 7 or $-5$

    **E.** 10 or $-12$

**32.** If $F(x) = x^2 + 5$ and $G(x) = 3x + 4$, which of the following is equivalent to $F(G(x))$?

    **A.** $x^3 + 4x^2 + 15x$

    **B.** $9x^2 + 24x + 21$

    **C.** $3x^2 + 19$

    **D.** $x^2 + 3x + 9$

    **E.** $x^3 + 20$

For questions 33 and 34, $A = \begin{bmatrix} 0 & 2 \\ 3 & 1 \end{bmatrix}$ and $B = \begin{bmatrix} 3 & 2 \\ 1 & 0 \end{bmatrix}$.

**33.** Which of the following is equivalent to $A - B$?

A. $\begin{bmatrix} 0 & 4 \\ 3 & 0 \end{bmatrix}$

B. $\begin{bmatrix} -3 & 0 \\ 2 & 1 \end{bmatrix}$

C. $\begin{bmatrix} 3 & 0 \\ -2 & -1 \end{bmatrix}$

D. $\begin{bmatrix} 3 & 4 \\ 4 & 1 \end{bmatrix}$

E. $\begin{bmatrix} -3 & -4 \\ -4 & -1 \end{bmatrix}$

**34.** Which of the following is equivalent to $AB$?

A. $\begin{bmatrix} 0 & 4 \\ 3 & 0 \end{bmatrix}$

B. $\begin{bmatrix} 2 & 0 \\ 10 & 6 \end{bmatrix}$

C. $\begin{bmatrix} 6 & 8 \\ 0 & 2 \end{bmatrix}$

D. $\begin{bmatrix} 1 & 0 \\ 0 & 1 \end{bmatrix}$

E. $\begin{bmatrix} -1 & 0 \\ 0 & -1 \end{bmatrix}$

**35.** What is (are) the value(s) for $x^2$ in the equation $\dfrac{8}{\sqrt{x^2 - 20}} = 2$?

A. 6

B. $-6$

C. $\pm 6$

D. 36

E. $\pm\sqrt{6}$

**36.**    If the area of an equilateral triangle is $100\sqrt{3}$, what is its perimeter?

A.    80

B.    60

C.    40

D.    30

E.    20

**37.**    Look at the following figure of two concentric circles, for which $\overline{PQ}$ is a radius of the inner circle and $\overline{PR}$ is a radius of the outer circle.

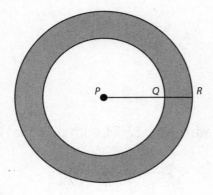

If $PQ = 6$ and $QR = 2$, what is the area of the shaded region?

A.    $28\pi$

B.    $32\pi$

C.    $36\pi$

D.    $48\pi$

E.    $64\pi$

**38.**    In the following figure, $AC = BD = 40$ and each of $\overset{\frown}{AB}$ and $\overset{\frown}{CD}$ is a semicircle whose diameter is 20.

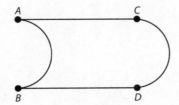

What is the perimeter of this figure?

A. $80 + 20\pi$

B. $80 + 40\pi$

C. $120 + 20\pi$

D. $120 + 40\pi$

E. $160\pi$

**39.** Refer to the figure of #38. What is its area?

A. $800 - 100\pi$

B. $800 - 50\pi$

C. $800$

D. $800 + 50\pi$

E. $800 + 100\pi$

**40.** Look at the following quadrilateral *EFGH*, with diagonal $\overline{EG}$.

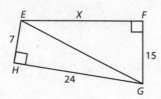

What is the value of *x*?

A. 16

B. 17

C. 20

D. 21

E. 22

**41.** Look at the following figure in which line $l_1$ is parallel to line $l_2$

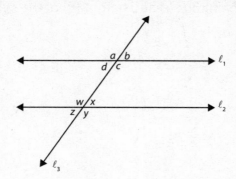

Which group of three angles have the same measure?

**A.**   *a, b, c*

**B.**   *a, c, w*

**C.**   *w, y, z*

**D.**   *a, c, z*

**E.**   *a, d, x*

**42.** Look at the following figure.

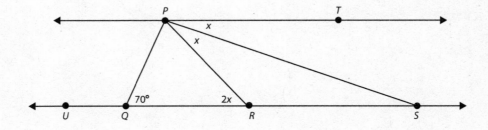

What is the value of *x*?

**A.**   30°

**B.**   35°

**C.**   40°

**D.**   45°

**E.**   Cannot be determined

**43.** A sphere and a cylinder have the same radius and the same volume. What is the relationship between the radius ($r$) and height ($h$) of the cylinder?

A. $r = h$

B. $r = \dfrac{4}{3}h$

C. $r = \dfrac{3}{4}h$

D. $r = \dfrac{1}{9}h$

E. $r = 9h$

**44.** If $\log_{\frac{1}{4}} x = -\dfrac{3}{2}$, what is the value of $x$?

A. $\dfrac{1}{8}$

B. $\dfrac{1}{4}$

C. $1$

D. $4$

E. $8$

**45.** If $\cos x = -\dfrac{1}{2}$ and $\sin x > 0$, what is the value of $x$?

A. $30°$

B. $60°$

C. $120°$

D. $240°$

E. $300°$

**46.**  Look at the following figure of a wire that is attached to the ground and to the top of a tree.

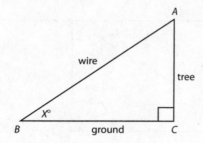

If $BC = s$, which of the following represents the length of the wire?

A.    $(s)(\cos x)$

B.    $(s)(\sin x)$

C.    $(s)(\tan x)$

D.    $\dfrac{s}{\sin x}$

E.    $\dfrac{s}{\cos x}$

**47.**  If $\cos A = \dfrac{3}{5}$ and $0° < A < 90°$, what is the value of $\sin(2A)$?

(Hint: Use the formula $\sin(2A) = 2 \sin A \cos A$.)

A.    $\dfrac{7}{25}$

B.    $\dfrac{17}{25}$

C.    $\dfrac{24}{25}$

D.    $1$

E.    $\dfrac{7}{5}$

**48.** Look at the following figure of a lighthouse and a ship, for which the angle of depression from the top of the lighthouse to the ship is represented as $y°$.

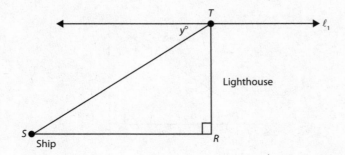

If the length of $\overline{RT}$ is represented as $L$, which of the following represents the distance from the ship to the base of the lighthouse?

A. $(L)(\tan y)$

B. $\dfrac{L}{\tan y}$

C. $\dfrac{L}{\sin y}$

D. $(L)(\sin y)$

E. $\dfrac{L}{\cos y}$

**49.** What is the smallest positive value of $x$, in degrees, such that $\sin(10x) = 0$?

A. 36

B. 27

C. 18

D. 9

E. 0

**50.**   What is the value of tan 330°?

   A.   $-\sqrt{3}$

   B.   $-1$

   C.   $-\dfrac{1}{\sqrt{3}}$

   D.   $\dfrac{1}{\sqrt{3}}$

   E.   $1$

## SOLUTIONS

**1.**   **E:**   $4 - 7 - (-7)^2 = 4 - 7 - 49 = -52.$

**2.**   **C:**   $\dfrac{4}{800} = 0.005 = 0.5\%.$  Be careful on this type of question!  Answer choice E in <u>wrong</u> because the question asks for a percent.

**3.**   **C:**   $\sqrt{\dfrac{8}{18}} = \sqrt{\dfrac{4}{9}} = \dfrac{2}{3} = 0.\overline{6}.$

**4.**   **E:**   Cross-multiply to get $7x - 49 = 3x - 9.$ Then $4x = 40,$ so $x = 10.$

**5.**   **B:**   The mean is the sum of the grades (980) divided by 10, which is 98. The median is the average of the fifth and sixth grades, which is $\dfrac{100+99}{2} = 99.5.$ The mode is the most frequent grade, which is 100.

The sum of these three statistical measures is 297.5.

**6.**   **D:**   $7{,}000{,}000 + 42{,}500 = 7{,}042{,}500 = 7.0425 \times 10^6.$

**7.**   **D:**   Let $x$ represent the number. Then $300 = 0.15\, x,$ so $x = \dfrac{300}{0.15} = 2{,}000.$

**8.**   **D:**   Following the rules of exponents, $6^{10} \times 6^{22} = 6^{10 + 22} = 6^{32}.$

**9.**   **B:**   The mean cost is $\dfrac{(12)(\$4)+(8)(\$10)}{20} = \dfrac{\$128}{20} = \$6.40.$

**10.** **E:** Factor the left side so that the equation reads as $(x)(x-2)(x+6)=0$. Then $x=0$ or $x-2=0$, or $x+6=0$. Thus, the roots are 0, 2, and $-6$.

**11.** **C:** Removing parentheses, we get $10a-5b-5b+2c=10a-10b+2c$.

**12.** **A:** $\sqrt{50x^5}+5x\sqrt{8x^3}=\sqrt{(25)(2)(x^4)(x)}+5x\sqrt{(4)(2)(x^2)(x)}=$

$5x^2\sqrt{2x}+(5x)(2x)\sqrt{2x}=15x^2\sqrt{2x}$.

**13.** **A:** The equation simplifies to $11x+11=0$. Then $x=\dfrac{-11}{11}=-1$.

**14.** **A:** Tom's age now is $X+Y$. So, in $Z$ years, he will be $X+Y+Z$ years old.

**15.** **C:** "Two less than three times $x$" becomes $3x-2$, and "four more than five times $x$" becomes $5x+4$.

**16.** **D:** $(3ab^2c^3)^2=(3^2)(a^{1\times2})(b^{2\times2})(c^{3\times2})=9a^2b^4c^6$.

**17.** **E:** If $|x-5|=|x+5|$, then either $x-5=x+5$ or $x-5=-(x+5)$. The first of these equations has no solution. The second equation can be written as $x-5=-x-5$, which simplifies to $2x=0$. Thus, $x=0$.

**18.** **C:** $\dfrac{x^2-4x+3}{x^2-6x+5}=\dfrac{(x-1)(x-3)}{(x-1)(x-5)}$. Cancel the common factor of $(x-1)$ in both

numerator and denominator to get $\dfrac{x-3}{x-5}$.

**19.** **A:** $\dfrac{(10a^4)^4(2a^6)^3}{16a^{20}}=\dfrac{(10,000a^{16})(8a^{18})}{16a^{20}}=5,000a^{16+18-20}=5,000a^{14}$.

**20.** **D:** The lowest common denominator is $18ab^2$. Then,

$\dfrac{10}{9ab^2}+\dfrac{5}{6b}=\dfrac{10}{9ab^2}\times\dfrac{2}{2}+\dfrac{5}{6b}\times\dfrac{3ab}{3ab}=\dfrac{20+15ab}{18ab^2}$.

**21.** **D:** $(4x^2-5y^3)^2=(4x^2)^2-(2)(4x^2)(5y^3)+(5y^3)^2=16x^4-40x^2y^3+25y^6$.

**22.** **A:** Rewrite $3x+4y=5$ as $y=-\dfrac{3}{4}x+\dfrac{5}{4}$. The slope of the line represented by this

equation is $-\dfrac{3}{4}$. Thus, the slope of a perpendicular line must be the negative

reciprocal of $-\dfrac{3}{4}$, which is $\dfrac{4}{3}$.

**23.** **B:** $4^{-\frac{3}{2}} - \dfrac{1}{125^{-\frac{1}{3}}} = \dfrac{1}{(\sqrt{4})^3} - \sqrt[3]{125} = \dfrac{1}{8} - 5 = -\dfrac{39}{8}.$

**24.** **D:** The required distance is $\sqrt{(\sqrt{2}+1-\sqrt{2})^2 + (\sqrt{2}-(\sqrt{2}-1))^2} =$

$\sqrt{1^2+1^2} = \sqrt{2}.$

**25.** **C:** $= \dfrac{a^{-4}b^{-5}c^8}{a^6 b^{-2} c^{-3}} = a^{-4-6} b^{-5+2} c^{8+3} = a^{-10} b^{-3} c^{11} = \dfrac{c^{11}}{a^{10}b^3}.$

**26.** **A:** $\dfrac{(x^4+x^2)^2}{x^4} = \dfrac{(x^4)^2 + 2(x^4)(x^2) + (x^2)^2}{x^4} = \dfrac{x^8+2x^6+x^4}{x^4} = x^4+2x^2+1.$

**27.** **A:** $\sqrt[3]{b^2} \times \sqrt[5]{b^7} = b^{\frac{2}{3}} \times b^{\frac{7}{5}} = b^{\frac{10}{15}+\frac{21}{15}} = b^{\frac{31}{15}} = \sqrt[15]{b^{31}}.$

**28.** **E:** Let $x$ represent the speed in miles per hour of the slower train and $x + 40$ represent the speed in miles per hour of the faster train.  The slower train traveled for six hours ( 12:00 noon to 6:00PM), whereas  the faster train traveled for  four hours (2:00PM to 6:00PM). Their distances were equal, so $(x)(6) = (x+40)(4)$. Then $6x = 4x + 160$, $2x = 160$, $x = 80$. The required distance is $(80)(6) = 480$ miles.

**29.** **D:** $(2 + 3i)(3 + i) = 6 + 2i + 9i + 3i^2$. Since $i^2 = -1$, the answer is $3 + 11i$.

**30.** **B:** The number of odd integers from 1 to 243 inclusive is equal to the number of even integers from 2 to 244 inclusive, which is $\dfrac{244}{2} = 122$. Since the odd integers 1, 3, 5, and 7 are missing from the given arithmetic progression, the number of terms must be $122 - 4 = 118$.

**31.** **D:** $\begin{vmatrix} x & 4 \\ 6 & x-2 \end{vmatrix} = (x)(x-2) - 24 = x^2 - 2x - 24$. Then $x^2 - 2x - 24 = 11$ can  be

written as $x^2 - 2x - 35 = 0$.  By factoring the left side, we get $(x - 7)(x + 5) = 0$. Thus, $x = 7$ or $x = -5$.

**32.** **B:** $F(G(x)) = F(3x + 4) = (3x + 4)^2 + 5 = 9x^2 + 12x + 12x + 16 + 5 = 9x^2 + 24x + 21.$

**33.**    **B:**    $A - B = \begin{bmatrix} 0-3 & 2-2 \\ 3-1 & 1-0 \end{bmatrix} = \begin{bmatrix} -3 & 0 \\ 2 & 1 \end{bmatrix}.$

**34.**    **B:**    $AB = \begin{bmatrix} (0)(3)+(2)(1) & (0)(2)+(2)(0) \\ (3)(3)+(1)(1) & (3)(2)+(1)(0) \end{bmatrix} = \begin{bmatrix} 2 & 0 \\ 10 & 6 \end{bmatrix}.$

**35.**    **D:**    Multiply both sides of the equation by the denominator to get $8 = 2\sqrt{x^2 - 20}$. Divide by 2 to get $4 = \sqrt{x^2 - 20}$. The next step is to square both sides to get 16 $= x^2 - 20$, so $x^2 = 36$. (Be sure that you read a question such as this very carefully! You are not asked to find the value of $x$.)

**36.**    **B:**    Let s represent one side. Then $\frac{s^2}{4}\sqrt{3} = 100\sqrt{3}$, which simplifies to $\frac{s^2}{4} = 100$.

So $s^2 = 400$, which means that s = 20. Thus, the perimeter is $(3)(20) = 60$.

**37.**    **A:**    The area of the shaded region is $(\pi)(8^2) - (\pi)(6^2) = 64\pi - 36\pi = 28\pi$.

**38.**    **A:**    The perimeter is the sum of two lengths of 40 each, plus the circumference of a circle with a diameter of 20. Thus, the perimeter is $40 + 40 + (\pi)(20) = 80 + 20\pi$.

**39.**    **C:**    The area of the figure is equivalent to the area of rectangle $ACDB$ minus the area of the semicircle that includes $\overset{\frown}{AB}$ plus the area of the semicircle that includes $\overset{\frown}{CD}$. Since the two semicircles have the same area, we only need the area of rectangle $ACDB$, which is $(40)(20) = 800$.

**40.**    **C:**    Use the Pythagorean theorem in triangle $EGH$. Then $EG^2 = 7^2 + 24^2 = 49 + 576 = 625$. So $EG = \sqrt{625} = 25$. Now apply the Pythagorean theorem to triangle $EFG$. Then $x^2 + 15^2 = 25^2$, which becomes $x^2 = 25^2 - 15^2 = 625 - 225 = 400$. Therefore, $x = \sqrt{400} = 20$.

**41.**    **B:**    $\angle a$ is congruent to $\angle c$ because they are vertical angles. $\angle c$ is congruent to $\angle w$ because they are alternate interior angles of parallel lines. Thus, $\angle a \cong \angle c \cong \angle w$.

**42.**    **E:**    The measure of each of $\angle TPR$ and $\angle PRQ$ is $2x$ degrees. Since they are congruent alternate interior angles, $\overrightarrow{PT}$ is parallel to $\overrightarrow{QS}$. However, we cannot find the measure of any other angle besides $\angle PQR$ and $\angle PQU$.

**43.**   **C:**    $\frac{4}{3}\pi r^3 = \pi r^2 h$. Now divide both sides by $\frac{4}{3}\pi r^2$, so that $r = \dfrac{\pi r^2 h}{\frac{4}{3}\pi r^2} = \dfrac{3}{4}h$.

**44.**   **E:**    $x = \left(\dfrac{1}{4}\right)^{\frac{-3}{2}} = 4^{\frac{3}{2}} = \left(\sqrt{4}\right)^3 = 2^3 = 8$.

**45.**   **C:**    Since $\cos x < 0$ and $\sin x > 0$, $x$ must lie in the second quadrant. Thus,

$$x = \cos^{-1}\left(-\frac{1}{2}\right) = 120°.$$

**46.**   **E:**    Let $w$ represent the length of the wire. By the definition of the cosine ratio, $\cos x = \dfrac{s}{w}$. Then multiply both sides by $w$ to get $(w)(\cos x) = s$. Now divide both sides by $\cos x$ to get $w = \dfrac{s}{\cos x}$.

**47.**   **C:**    Using the Pythagorean theorem, the side opposite $\angle A$ is $\sqrt{5^2 - 3^2} = \sqrt{25 - 9} = \sqrt{16} = 4$ Then $\sin A = \dfrac{4}{5}$. Finally, $\sin(2A) = (2)(\dfrac{4}{5})(\dfrac{3}{5}) = \dfrac{24}{25}$.

**48.**   **B:**    Line $l_1$ and $\overline{RS}$ are parallel to each other. So the measure of $\angle S$ is $y°$ because they are alternate interior angles. Then $\tan y = \dfrac{L}{RS}$. Multiply both sides by $RS$ to get $(RS)(\tan y) = L$. Thus, $RS = \dfrac{L}{\tan y}$.

**49.**   **C:**    If the inverse sine of an angle is 0, the angle must be 0° or an integral multiple of 180°. This means that either $10x = 0°$ or a multiple of 180°. If $10x = 0°$, then $x = 0°$. However, we need the smallest positive number and zero is not a positive number. Thus, the smallest positive number is the solution to $10x = 180°$, which is $x = 18°$.

**50.**   **C:**    The reference angle for 330° is 360° − 330° = 30°. Since 330° lies in the fourth quadrant, the tangent value is negative. Thus, $\tan 330° = -\tan 30° = -\dfrac{1}{\sqrt{3}}$.

# COMPASS – PRACTICE TEST 2 ANSWER KEY

| Question Number | Answer Key | Content Domain | Question Number | Answer Key | Content Domain |
|---|---|---|---|---|---|
| 1 | E | Pre-Algebra | 26 | A | Algebra |
| 2 | C | Pre-Algebra | 27 | A | College Algebra |
| 3 | C | Pre-Algebra | 28 | E | Algebra |
| 4 | E | Algebra | 29 | D | College Algebra |
| 5 | B | Pre-Algebra | 30 | B | College Algebra |
| 6 | D | Pre-Algebra | 31 | D | College Algebra |
| 7 | D | Pre-Algebra | 32 | B | College Algebra |
| 8 | D | Pre-Algebra | 33 | B | College Algebra |
| 9 | B | Pre-Algebra | 34 | B | College Algebra |
| 10 | E | Algebra | 35 | D | College Algebra |
| 11 | C | Algebra | 36 | B | Geometry |
| 12 | A | College Algebra | 37 | A | Geometry |
| 13 | A | Algebra | 38 | A | Geometry |
| 14 | A | Algebra | 39 | C | Geometry |
| 15 | C | Algebra | 40 | C | Geometry |
| 16 | D | Algebra | 41 | B | Geometry |
| 17 | E | Algebra | 42 | E | Geometry |
| 18 | C | Algebra | 43 | C | Geometry |
| 19 | A | Algebra | 44 | E | College Algebra |
| 20 | D | Algebra | 45 | C | Trigonometry |
| 21 | D | Algebra | 46 | E | Trigonometry |
| 22 | A | Algebra | 47 | C | Trigonometry |
| 23 | B | College Algebra | 48 | B | Trigonometry |
| 24 | D | Algebra | 49 | C | Trigonometry |
| 25 | C | College Algebra | 50 | C | Trigonometry |

# INDEX